MICROSCOPY HANDBOOKS 12

An introduction t[...] acoustic microscop[...]

Andrew Briggs
Department of Metallurgy and Science of Materials,
University of Oxford

Oxford University Press · Royal Microscopical Society · 1985

Oxford University Press, Walton Street, Oxford OX2 6DP

Oxford New York Toronto
Delhi Bombay Calcutta Madras Karachi
Kuala Lumpur Singapore Hong Kong Tokyo
Nairobi Dar es Salaam Cape Town
Melbourne Auckland

and associated companies in
Beirut Berlin Ibadan Nicosia

Royal Microscopical Society
37/38 St. Clements
Oxford OX4 1AJ

Published in the United States
by Oxford University Press, New York

British Library Cataloguing in Publication Data
Briggs, Andrew
An introduction to scanning acoustic microscopy.—(Microscopy handbooks; 12)
1. Sound—Measurement—Equipment and supplies
2. Microscope and microscopy
I. Title II. Series
534'.42 QC228.3

ISBN 0-19-856412-0

Library of Congress Cataloging in Publication Data
Briggs, Andrew.
Introduction to acoustic microscopy.
(Microscopy handbooks; 12)
Bibliography: p.
Includes index.
1. Materials—Microscopy. 2. Acoustic microscopes.
I. Title. II. Series.
TA417.23.B75 1985 620.1'127 85-4987
ISBN 0-19-856412-0 (pbk.)

Typeset and printed in Great Britain by Alden Press, Oxford

Contents

Introduction

The significance of the scanning acoustic microscope does not lie in its resolution alone. Even the spectacular resolution of 20 nm that has been obtained in a cryogenic acoustic microscope (Hadimioglu and Foster 1984) can easily be surpassed by electron microscopes. And while the submicron resolution that can be obtained with the water-coupled instruments to be described in this book represents an outstanding achievement of innovative research, comparable resolution can still be obtained in an optical microscope with a great deal less trouble and expense. There are two advantages in using acoustic waves for producing images. The first lies in the ability of ultrasonic waves to penetrate materials that are opaque to other kinds of radiation, most notably light. The second lies in the distinctive origin of the contrast in the mechanical properties of the specimen. The first of these advantages has been exploited in the main uses of ultrasound that have been established so far. The use of ultrasonic waves was first developed for the detection of submarines in the first world war (for a fascinating historical account of this see Graff 1981). Since then two areas that have also been extensively developed are medical imaging and non-destructive testing. In medical imaging use is made of the fact that ultrasonic waves can penetrate through body tissue, and can be weakly scattered by changes in the density or elasticity of the tissue. The reflected echoes are detected and can be used to build up a scanned image. In the early days the image was built up by manually scanning a transducer, and each echo was displayed on a point on the image that corresponded to the direction in which the transducer was pointing at that moment and also to the time taken for the echo to return. More recent instruments use an array of transducers that are fixed mechanically, and the scanning (or beam steering) is performed electronically by means of delays and phase shifts introduced into the signals to and from each individual transducer. Whichever method of scanning is used, the information is displayed in the form of an image, usually in the form of a B-scan tomogram, which corresponds to an imaginary slice through the body in the plane in which the ultrasonic waves are scanned. In ultrasonic non-destructive testing (n.d.t.) the ability of ultrasonic waves to penetrate opaque materials is again used, usually through steels, though other alloys and even ceramics and other materials are also inspected. In this case the echoes can be much stronger, because reflections usually occur at cracks or voids, and almost 100 per cent of the signal may be scattered, though not necessarily back to the receiving transducer. In conventional ultrasonic non-destructive testing, inspection is made only to discover the presence of defects; for this purpose the A-scan (the simple display of signal amplitude against time on an oscilloscope) alone is used, and no attempt is made to obtain an image. More recent developments have enabled good images to be formed. The most important of these is the time-of-flight method, in which the time taken for echoes diffracted at the edges of defects to arrive at the receiving transducer are measured, and by suitable signal processing the position of the defects is deduced. Such scattered signals may be weak, but signal-to-noise ratio may be improved by signal averaging. The information obtained by the time-of-flight method can then be displayed as a suitable image. In inspecting a structure, such as a component of a nuclear reactor, for defects, an array of transducers may be scanned over the surface of the component. Thus data may be collected from the whole of the component, and images of suitable sections may be produced. In this way it is possible to measure the size and orientation of defects, and by using fracture mechanics and knowledge of the stresses in service, fitness for purpose can be assessed.

In both medical imaging and n.d.t. the ultrasonic frequency range typically used is from 2 to 10 MHz. At 5 MHz acoustic waves have a wavelength of about 0.3 mm in water and 1.2 mm in steel. The resolution obtained by these methods is therefore of that order for high contrast structures. Since the resolution varies inversely with frequency, by using acoustic waves of frequency in the GHz range is is possible to obtain wavelengths of the order of the wavelength of light, and therefore to obtain resolution comparable to that of an optical microscope. This point was appreciated as early as 1936, but the technology of the time did not allow the idea to be implemented then. There have been a number of attempts to make an acoustic microscope since then, but many of them were of only limited success because they involved trying to focus the acoustic image onto a plane, and then visualize it by converting the acoustic field into either a visual image or into an electrical pattern (a brief historical survey of the development of acoustic microscopy may be found in the introduction to Lemons and Quate 1979). Three kinds of acoustic microscopy have proved successful: scanning laser acoustic microscopy, scanning electron acoustic microscopy, and the subject of this book, which has become known as scanning acoustic microscopy, or sometimes simply as acoustic microscopy. In scanning laser acoustic microscopy, ultrasonic waves are transmitted into the whole of the specimen, and the resulting disturbance of the surface is measured point by point by a laser beam that is scanned over the surface. This technique is described and compared with scanning acoustic microscopy by Kessler and Yuhas (1979). In the scanning electron acoustic microscope a pulsed electron beam generates sound (usually by oscillating thermal expansion, though other mechanisms may also operate) point by point on the surface of the specimen, and sound waves from the whole volume are detected by a transducer in contact with the specimen. For a comparison of this technique with scanning acoustic microscopy see Briggs (1984). Thus in the scanning laser acoustic microscope, it is the detection of the sound that is scanned, and in the scanning electron acoustic microscope the source is scanned. In the scanning acoustic microscope, acoustic waves only are used, and both the source and the detector are scanned relative to the specimen, or alternatively they are fixed and the specimen is scanned relative to them.

The first scanning acoustic microscope operated in transmission (Lemons and Quate 1974). An acoustic wave was brought to a focus by a sapphire lens, and in the absence of a specimen this was received by an identical lens arranged confocally. The object was then introduced into the common focal zone of the two lenses, and scanned between them. In this mode it was not necessary to use pulsed signals, and simple continuous-wave electronics could be used. The first pictures were of biological specimens, and in imaging these it was not the ability of ultrasonic waves to penetrate opaque materials that was particularly important (since thin biological specimens are translucent), but the origin of the contrast in the mechanical properties of the specimens. Indeed, the advantage of the acoustic microscope lay in the fact that, whereas light is so little affected by thin sections of biological material that staining is necessary to render the features of interest visible, acoustic waves are sufficiently scattered by the variations in elastic properties that good contrast is obtained without the need for any kind of staining. It was only much later that the transmission microscope began to be used for imaging through opaque materials, for example to observe the development of a diffusion bond between two metal surfaces (Derby, Briggs, and Wallach 1983).

Gradually interest began to shift from transmission microscopy to reflection microscopy, with a single lens being used for both transmitting and receiving the acoustic signal (which has now to be pulsed). There were various reasons for this. Partly, no doubt, it was because of the shear mechanical difficulty, as the resolution began to be pushed higher and higher, of aligning the transmitting and receiving lens (the reflection microscope, of course is self-aligning provided the specimen is level). But while this is difficult it is by no means impossible. A more important reason may have been that interest began to grow in imaging specimens of solid materials. Although acoustic waves have the advantage of being able to propagate through materials that are opaque to light, they suffer from the disadvantage that unlike light, which can travel through a vacuum with zero attenuation, and through air with very little loss, acoustic waves always suffer attenuation in any medium. This attenuation can be quite low in a few solids such as sapphire (which is why sapphire is generally used for high frequency lenses), but in most fluids it rises as the square of the frequency. This means that for high resolution microscopy the fluid path between the lens and the specimen must be small, and for practical lens design this in turn means that the focal length of the lens is also small. Therefore specimens to be imaged in transmission must be thin. For biological specimens this often presents few problems, because techniques for making thin biological specimens are well established. For solids, however, the situation is quite different. Geologists are used to making thin specimens of rock for transmission optical microscopy, but for metallography it is very much easier to polish one surface only of the specimen, and with some other specimens, such as semiconductor devices, it is desirable to examine them as they are without any polishing at all.

It is therefore in the reflection mode that most of the recent development and application of the scanning acoustic microscope has occurred. All of the commercial scanning acoustic microscopes that are available are reflection instruments, and it is this kind of reflection scanning acoustic microscope and its uses that will be described in this book. A general view of a complete system, and the specimen stage and scanner of another instrument are shown inside the back cover. In the reflection mode the surface of the specimen is examined, and therefore the ability of acoustic waves to penetrate opaque materials is not exploited in the obvious sense. Rather it is the fact that the acoustic image contains information about the way that acoustic waves interact with the properties of the specimen that is of prime importance. And yet this itself depends on the ability of the specimen to support elastic waves. As will become clear, the excitation and propagation of surface waves, often referred to as Rayleigh waves, play a vital role in the generation of contrast in the acoustic microscope, and factors that affect these waves affect the contrast. Moreover, since Rayleigh waves extend about a wavelength below the surface, even in reflection it may be possible to image features that lie below an opaque surface layer.

The handbook is divided into two parts. The first part is concerned with the instrument. In Chapter 1 the principles of acoustic microscopy are outlined. This chapter is intended for a complete newcomer to the subject who wishes to discover how a scanning acoustic microscope works. Until now anyone undertaking research involving scanning acoustic microscopy had first to build his own instrument, or at least to join a laboratory where one was already operational, but now it is possible to buy a fully operational instrument, and Chapter 1 will help to give some appreciation of what has gone into the design and construction of such a microscope. Chapter 2 is very down-to-earth and practical, and is intended for the new owner of a microscope, or for someone from a different background who is about to start using a scanning acoustic microscope in his research. It contains practical help, much of which has been learned the hard way. Much of what appears in this chapter is the kind of detail that never appears in published papers, but which anyone with two or three years experience of using a microscope would know almost without thinking about it. It is hoped that this may prove a useful addition to what will be found in individual manufacturers' handbooks. The third chapter contains the elements of the contrast theory of the acoustic microscope. This depends on the variation of the signal, V, with the defocus, z, and has become known as the acoustic material signature, or simply $V(z)$. If Chapter 2 contains information of which there has been a dearth in the literature, Chapter 3 contains material of which there has been a glut, and it is hoped that in Chapter 3 it is presented in an accessable form.

A reader who wishes to know what the scanning acoustic microscope can do may wish to go straight to Part II, and to read about the images that have been obtained in specimens most closely corresponding to his own. For convenience these have been divided into four categories. In Chapter 4 contrast arising simply from changes in elastic properties is described. After briefly mentioning low resolution subsurface imaging of bonds, examples are given that most directly illustrate applications of the contrast theory presented in Chapter 3. These include composite materials, voids in superalloys, and regions of demineralization in human teeth. One of the simplest properties to image in the acoustic microscope is grain structure; this can be revealed in a polished polycrystalline specimen without etching. The theory of this is summarized in Chapter 5, and used to account for the fact that while poor images of grains are obtained from aluminium and copper, for different reasons in each case, excellent images are obtained from, for example, nickel. Chapter 6 considers effects that can perturb the propagation of elastic waves in the surface. These may be thin layers that affect the velocity of surface waves; this is generally the basis of the contrast that is obtained from integrated circuits, and enables, for example, the extent of adhesion of a surface film to be imaged. Alternatively, mechanisms that cause attenuation of elastic waves, such as scattering from grain boundaries, or damping by dislocations, can also affect $V(z)$, and therefore be the source of contrast for imaging properties that might otherwise not be visible. Some of the most dramatic results obtained with the acoustic microscope have been images of fine surface cracks. These may be much less than a wavelength wide, so that they would not be resolved according to conventional criteria, and yet they can show up with very strong contrast. Similar contrast can be obtained from interfaces between dissimilar materials and from grain boundaries. Such images are presented in Chapter 7. The contrast arises from the excitation of Rayleigh waves in the surface of the specimen that can be strongly scattered because they can strike the discontinuity broadside on. Indeed, the excitation and detection of leaky Rayleigh waves plays a dominant role in almost all the images presented in this book, and this point is drawn out in the conclusion.

Acknowledgements

Many people have contributed to this book, and I wish especially to thank my colleagues Mr J. M. R. Weaver, Dr C. Illett, Dr M. G. Somekh and Professor W. D. O'Brien for all that I have learnt from them and especially for the many results of theirs that appear. The development of acoustic microscopy at Oxford is in collaboration with AERE Harewell, and I am particularly grateful to Mr D. W. Evans, Mr R. Martin and Dr B. J. Smith for extensive development of the electronic system of our microscope. I also wish to thank Professor E. A. Ash and his colleagues at U.C.L. for much that I have learnt from them, and for incalculable help that they have given us in the supply of lenses. I am grateful to Mr R. D. Weglein, Dr A. J. Miller, Dr M. Nikoonahad, and Professor C. F. Quate for supplying figures. Anyone in this field owes an enormous debt to Professor Quate; it is a pleasure to acknowledge this and also to thank him for more personal help and encouragement. Dr C. Hammond, the editor of this series of handbooks, has suggested many helpful changes to the text. I am deeply grateful to the Royal Society for a Research Fellowship in the Physical Sciences. Finally, I wish to thank Professor Sir Peter Hirsch, who first introduced me to acoustic microscopy, for constant guidance and support.

Part I: The Instrument

1

Principles of scanning acoustic microscopy

The key to the successful development of scanning acoustic microscopy lay in the realization that, while it has not proved possible to form an extended image simultaneously of a whole object, it is possible to make an acoustic lens that has good focusing properties on axis (Lemons and Quate 1979). Such a lens can be used to focus acoustic waves onto a spot on a specimen, and also used to receive the acoustic energy from that spot. If the lens is then scanned over the specimen systematically, and the intensity of the reflected signal is sent to a synchronous display, a scanned image can be built up in a manner similar to scanning electron microscopy, or indeed to domestic television.

The heart of the acoustic microscope is therefore the lens; this is illustrated schematically in Fig. 1.1. The lens consists of a disc of sapphire oriented with its *c*-axis as the axis of the disc. One face of the disc is polished to be flat. The other

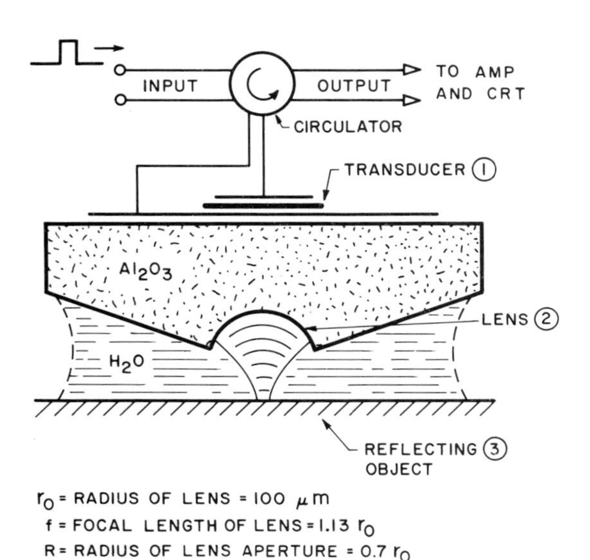

Fig. 1.1. The acoustic transducer and lens in the scanning acoustic microscope (Lemons and Quate 1979).

face has a concave spherical surface ground in it. A coupling fluid is necessary between the lens and the specimen; water is most usually used. The velocity of sound along the c-axis in sapphire is 11 100 m s^{-1} and in water it is about 1500 m s^{-1}. This means that there is a very large refractive index ($n = 7.4$) for acoustic waves striking the lens. If the refractive index were infinite, then plane waves crossing the sapphire–water interface would be refracted into spherical waves converging at the centre of curvature of the lens, or to put it another way rays parallel to the axis of the lens would be refracted along radii of the spherical surface. Of course the refractive index is not infinite, but it is much greater than would be encountered in any comparable optical system. The consequence of this is that a single lens surface is adequate even for a lens of large numerical aperture; at the wavelengths employed in acoustic microscopes the spherical aberrations are usually less than a tenth of a wavelength, and may therefore be neglected in comparison with diffraction limitations.

If the velocity mismatch between sapphire and water is large and is a good thing, the impedance mismatch is also large and is a bad thing. Acoustic impedance, Z, describes the ratio of traction to particle displacement velocity, and is equal to the product of acoustic velocity and density; the SI unit is a rayl: 1 rayl = 1 kg m^{-2} s^{-1}. The impedance of sapphire is 44.3 Mrayl, that of water is 1.5 Mrayl. The power transmitted across an impedance mismatch is $Z_1Z_2/(Z_1 - Z_2)^2$; for a sapphire–water interface at normal incidence this is 3.6 per cent, which means that for a wave passing back through the same interface again, only 0.13 per cent of the power would return from the specimen, even if there were no other losses. It is therefore necessary to apply a quarter-wave antireflection coating to the lens; this is analogous to the blooming on an optical lens, except that in the acoustic case the benefits are much greater. The optimum impedance for a quarter-wave impedance matching layer is given by $Z^2 = Z_1Z_2$; the ideal material for this is chalcogenide glass, whose composition can be adjusted to give exactly the right impedance (Kushibiki, Maehara, and Chubachi 1981). An acceptable substitute is SiO_2 (Kushibiki, Sannomiya and Chubachi, 1980).

The highest frequency at which a given lens can operate is determined by the radius of curvature and the attenuation of the coupling fluid. Most fluids at or near room temperature exhibit linear viscosity; this causes the attenuation of acoustic waves propagating through them to be proportional to the frequency squared. Thus for example the attenuation in water at 30° C is $\alpha\nu^2$, where $\alpha = 1.6 \times 10^{-13}$ dB m^{-1} Hz^{-2}, this means that in propagating a distance x the intensity would decrease by $10^{-0.1\alpha\nu^2 x}$. This provides a very sharp upper cutoff frequency above which a given lens cannot be operated. For example, at a frequency of 1 GHz (10^9 Hz) the attenuation is 166 dB mm^{-1}, so that the total loss in a lens of focal length 100 μm would be 33 dB, i.e. about 0.05 per cent of the transmitted power would be returned from the specimen. This would be about the greatest loss that one would wish to design for. Now supposing that the frequency were doubled: the attenuation would quadruple to 664 dB mm^{-1} and now the loss would be 133 dB, i.e. the return echo would be only 5×10^{-14} of the original intensity. Thus by merely trying to double the frequency the signal strength would have been reduced by a factor of 10^{10}. This would be far too small to detect, and the microscope would no longer function.

To increase the frequency, therefore, it is necessary to reduce the lens radius. Curiously enough this is not difficult; it has proved possible to grind radii as small as have ever been required. The problem is that it is never possible to eliminate the reflection from the lens surface completely, and as this signal does not experience the attenuation of the fluid path, it is generally larger than the reflection from the specimen. By operating the lens in a pulsed mode, it is possible to separate the echoes from the lens and those from the specimen because they arrive at different times. However as the focal length of the lens is made shorter, so the time interval between the arrival of the two echoes decreases. The shortest pulses of acoustic waves that can be made are about 20 ns (20×10^{-9} s) long, and with a velocity in water of 1.5 μm ns^{-1} this means that for the pulses to be well separated the lens must have a radius of about 30 μm, otherwise the specimen echo would be swamped by the lens echo. This means that the highest frequency for a microscope with water coupling at 30° C is about 2 GHz. The attenuation of acoustic waves in water decreases with increasing temperature; the temperature dependence of the attenuation and also of the velocity are plotted in Fig. 1.2. By raising the temperature of the water it is possible to reduce α by a factor of about 2, and this raises the operating frequency to about 3 GHz. By working very hard at making the pulses shorter it is possible to allow the focal length to be shorter and thus increase the frequency a little (Hadimioglu and Quate 1983), but these figures are representative of most water-coupled microscopes.

Since the spherical aberrations are negligible, the resolution is determined almost solely by diffraction limitations. The numerical aperture can be about 1, and for a simple microscope this would give a resolution of 1.22λ. In fact the resolution is rather better than this for two reasons. The first is that the acoustic microscope is a confocal system, which means that focusing occurs both when the acoustic waves

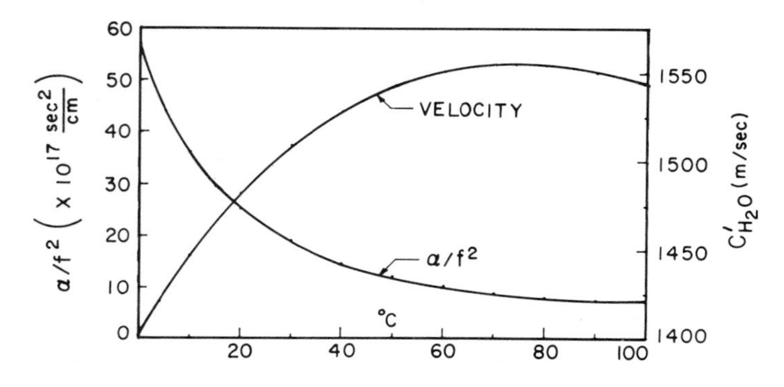

Fig. 1.2. Temperature dependence of acoustic velocity and attenuation in water (Lemons and Quate 1979).

travel towards the specimen and when they are detected again by the lens. Secondly, for a scanned image system it is possible to amplify any dip in brightness between the images of two point objects electronically, this means that the Rayleigh criterion is less relevant than the Sparrow criterion, which defines the limit of resolution as occurring when the dip has just flattened out, so that if the points were any closer there would be only one bump instead of two. The effect of these two factors is that the resolution of the acoustic microscope may actually be better than a wavelength; it is given by λ/N, where N is the numerical aperture (Kino 1980). Thus for a numerical aperture of one a microscope operating at 3 GHz would have a resolution of 0.5 μm. It is a fascinating coincidence that, for completely unrelated reasons, the limit of resolution in a water-coupled acoustic microscope is quite comparable to that obtainable in a light microscope (Quate 1980).

On the flat surface of the lens opposite the face with a spherical surface ground in it there is a transducer. For frequencies above 200 MHz the transducer material is ZnO, a strongly piezoelectric semiconductor. An epitaxial layer is grown by sputtering ZnO onto a gold film previously deposited on the sapphire. For lower frequencies thin plate transducers of $LiNbO_3$ or PZT can be bonded onto the lens instead of ZnO; also attenuation in the lens ceases to be a problem and fused quartz can be used instead of sapphire, with the advantage that its impedance is 13.1 Mrayl, which is much lower than that of sapphire, and therefore it is not necessary to use an antireflection coating.

A radio frequency (r.f.) signal of appropriate frequency, and about 200 mW power, is fed to a PIN switch, a fast solid-state switch that can give pulses of a few ns duration. This pulse enters one port of a single-pole-double-throw (s.p.d.t.) PIN switch whose common port is connected to the lens. The pulse energizes the transducer, which generates an acoustic pulse of roughly the same duration and frequency. The r.f. electrical switch from the returned acoustic echoes is directed by the s.p.d.t. switch to a low-noise high-grain receiving amplifier. One of the important properties of this amplifier is that it must be able to recover very quickly from overloads. The s.p.d.t. switch has at best an isolation of 60 dB, and this means that the breakthrough from the transmitting r.f. pulse can saturate the amplifier. Moreover, PIN switches themselves can give out spurious spikes, and these can also cause saturation. Some microscopes use a circulator in place of an s.p.d.t. switch; this is a passive device rather like a roundabout that makes the r.f. signal circulate in a given sense and take the next exit. Whichever method is used, the reflection from the lens surface will also be large, and the important point is that by the time that the specimen reflection arrives at the transducer, the amplifier must be in a state to amplify the specimen echo. The output of the r.f. amplifier is fed to a diode detector and then to a pulse amplifier of sufficient bandwidth to amplify the envelope of the r.f. signals without distortion. At this stage the signal may be displayed on an oscilloscope, as shown in Fig. 1.3. The first pulse consists of breakthrough from the transmitted pulse, plus spikes from the PIN switches. The second pulse is the acoustic reflection from the lens. The fourth pulse is another echo from the lens surface that has reverberated once inside the sapphire. The third pulse is

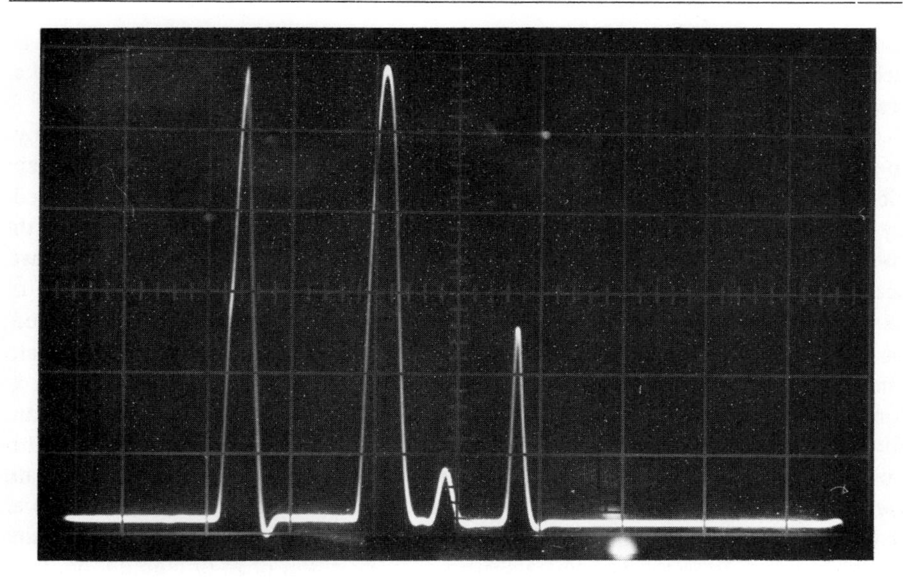

Fig. 1.3. Oscilloscope display of the detected signals from the lens.

the signal of interest; this is the echo from the specimen itself. It can be seen that this is smaller than any of the other pulses (which may indeed be saturating the amplifiers); this is why it is such an important part of the design of the system to ensure that this pulse can be separated from the others (including the second reverberation), and also that the amplifiers have recovered from any saturation.

The intensity of the echo from the specimen is measured by a fast circuit that measures the height of a chosen pulse and holds this value as a d.c. signal until the next r.f. pulse is transmitted. This signal is passed on to a frame store which holds that value for intensity at an address that corresponds to the position of the lens relative to the specimen. The lens is then scanned in a raster pattern, with the intensity of an echo being recorded at each position and in this way a complete image is built up. Simple considerations show the minimum time that this requires. The timescale in Fig. 1.3 is 200 ns per division. The full sequence of echoes is not shown (it is cut off by the s.p.d.t. switch), but typically at least 10 μs should be left for the reverberations to die down before launching a new acoustic pulse. This means that the maximum pulse repetition frequency is about 100 kHz. A digital framestore usually stores 512 × 512 pixels so this would require 2.6 s if pixels were being recorded at a constant rate. But in practice it is not wise to record in both directions of scan (because it is almost impossible to avoid a double image resulting) and also time must be allowed for turnround at the ends of the scan. This means that a good resolution picture takes at least about 10 s to record. The lens that produced the echoes of Fig. 1.3 has a radius of 120 μm; a smaller lens would allow a reduction in scanning time, though 50 Hz is about the fastest rate that is used in practice. For lower frequency lenses a correspondingly lower scanning rate is necessary, though in this case the limitation generally lies in the rate at which the lens can be scanned over a reasonable distance.

Scanning in the fast (X) direction is achieved in high resolution microscopes by mounting the lens on some kind of leaf spring arrangement and driving it with an electromagnetic vibrator. This is driven from a medium power amplifier energized by a suitable oscillator. The framestore may be controlled either in open loop, with the address of each pixel determined by a sweep that is synchronized to the fast scan driver, or in closed loop, with the position of the lens being measured (for example by an l.v.d.t.) and then fed to the framestore. The line scan can be displayed on a c.r.t. with the vertical displacement corresponding to the peak detector signal, and the horizontal displacement to the lens position. For comfortable viewing a long persistence phosphor is used, with blanking of the reverse scan; The line scan display does not form part of the eventual imaging system, but it is invaluable for optimizing the adjustment of the microscope for each specimen. The slow (Y) scan can be either another leaf spring arrangement driven in the same way, or else a linear translation stage driven by either a stepper motor or a d.c. motor. Once again the position information to the framestore may be either open or closed loop. For very low magnifications it is difficult to make leaf springs with large enough scans, and then stepper motors are used for both directions of scan.

It is a semantic question where the boundary lies between what in n.d.t. or medical imaging would be called a C-scan and acoustic microscopy. For the present purpose acoustic microscopy will be regarded as being above about 50 MHz. All reflection acoustic imaging above this frequency has various features in common. The acoustic waves are brought to a diffraction limited focus by a single lens surface, and to form an image the lens is mechanically scanned relative to the specimen. A coupling fluid must be used, and usually this is water. The limit of resolution in an instrument operating near room temperature is about the same as that of a light microscope. The only way to improve on this resolution is to find a coupling fluid that has a much lower attenuation. The only candidate that gives improvement by more than a factor of two or so is liquid helium; below about 0.2 K the attenuation becomes negligible, and some spectacular results have been obtained with a resolution of 20 nm (Hadimioglu and Foster 1984). However for most purposes the room temperature water-coupled instruments are to be preferred, and we now turn to the practical operation of these.

2

Practice of scanning acoustic microscopy

In order to understand what is important in obtaining good images from a high resolution acoustic microscope, it is essential to grasp the effect of defocus on contrast. The theory of this will be considered in the next chapter; for this chapter it is merely necessary to understand the experimental aspect (Weglein and Wilson 1978). If the X and Y scans (i.e. those parallel to the lens surfaces) are switched off and instead the specimen is scanned in the Z direction (i.e. along the lens axis), then the resulting variation of transducer signal, V, as a function of defocus, z, is known as $V(z)$ (Quate, Atlar, and Wickramasinghe 1979). By convention $z = 0$ refers to the focal plane of the lens, and moving the specimen closer to the lens (which is where most of the interesting phenomena occur) corresponds to $-z$. An experimental $V(z)$ curve for an aluminium specimen is shown in Fig. 2.1; this was obtained at 0.2 GHz, at which frequency the wavelength in water is 7.5 μm (most of the effects described here scale with wavelength). The first feature to notice about this $V(z)$ curve is that the depth of focus of the acoustic microscope is rather small. Thus relative to focus, if the specimen is moved more than about 20 μm in either direction the signal falls to less than half its value. Beyond this, especially in the negative z regime, $V(z)$ oscillates through a series of peaks and troughs before eventually dying away (if the specimen has not already met the lens). Thus if the value of z were to vary by more than a few μm across a specimen that was being imaged, there would be a series of dark fringes across the picture. These would correspond to the minima in $V(z)$, and would follow contours in the surface height if the specimen were not flat, or would form a set of parallel fringes if it were flat but not level.

Actually the situation is rather worse than this for at least two reasons. The first is that very little contrast is usually found by operating the microscope at focus. Two regions of a specimen with different elastic properties will often give similar $V(z)$ curves at and beyond $z = 0$. The differences in $V(z)$ usually begin to manifest themselves in the slope of $V(z)$ as the specimen is moved towards the lens, i.e. for negative values of z. Second, even when z is chosen to maximize contrast between regions of different properties, the differences may be only a small fraction of the average intensity. In that case the brightness and contrast controls of the framestore or monitor may be adjusted to allow these small variations to fill the whole dynamic range of the display system. But these two together mean that the microscope may be imaging over a part of the $V(z)$ curve where the slope is very steep, with the controls set to go from black to white for a change of only a small fraction of the

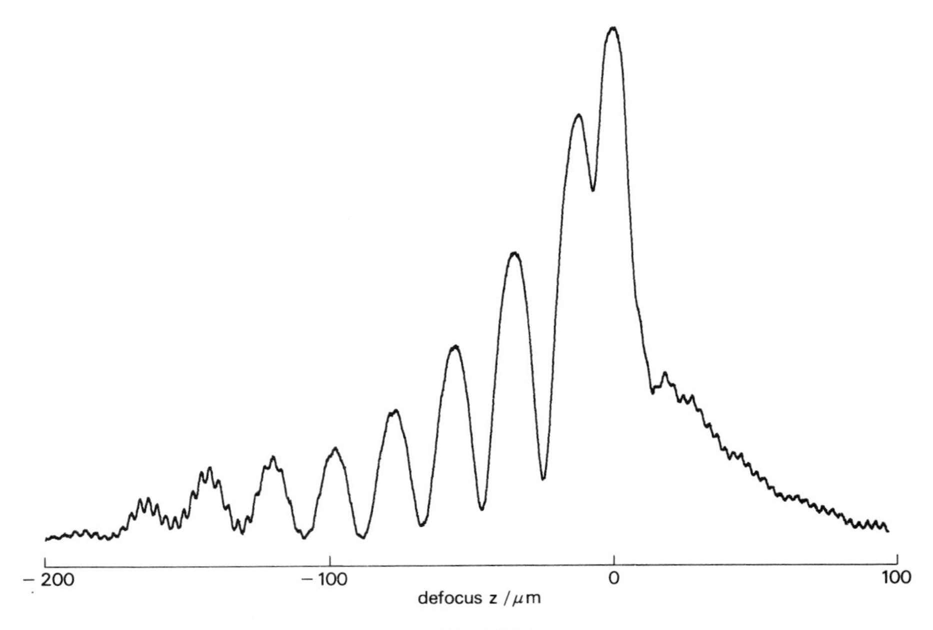

Fig. 2.1. Experimental $V(z)$ for aluminium at 237.5 MHz.

signal intensity. There is a third possible factor which may make the problem even more difficult in some microscopes: the measured $V(z)$ curve may have a fine structure superimposed on the main peaks and troughs. Ways of reducing this will be discussed below, but in so far as this makes the gradient of $V(z)$ even steeper over the range of z used for imaging, and have oscillations that are more closely spaced, it increases still further the sensitivity of the image to variations in z. For all these reasons the greatest priority in specimen preparation is to get the specimen flat and level.

The acoustic microscope is more sensitive to surface topography than to any other specimen characteristic. However this is not usually of great interest; topography can be imaged in a light microscope with far less trouble, and in an s.e.m. with far higher resolution. Therefore if topography is not to mask contrast from elastic properties, the specimen must be prepared with its surface flat and smooth to better than a wavelength. Three categories of specimens may be identified. The first includes those that already have the desired surface finish, such as semiconductor devices and thin films deposited on previously polished surfaces; these need no further preparation except cleaning, though the effect of surface steps on an integrated circuit should be remembered when interpreting images. The second includes most other materials, and these must be suitably prepared. With materials of fairly uniform hardness this amounts to preparing a good metallographic finish. This can be achieved by grinding on carborundum paper of successively finer grades down to 1200, followed by polishing successively with 6, 3, and 1 μm diamond paste. This has proved quite satisfactory for microscopy at 0.75 GHz; for higher frequencies the final diamond size should be scaled with the wavelength. Care should be taken not to polish for any longer than necessary on each size of diamond, as excess polishing tends to reduce flatness. A better way to obtain the desired finish is to use a good lapping machine. In this the specimen can be lapped first with 6 μm paste on a copper wheel, and then with 1 μm paste on a solder wheel (again the final diamond size scales with wavelength). By this technique excellent results may be reliably obtained, with a high degree of smoothness and flatness over a large area of the specimen, and for specimens of variable hardness differential polishing is greatly reduced. Etching should not be used; grain structure can be seen without it, and it may mask elastic properties of greater interest. The third category is of soft biological specimens. Although histological preparation for light microscopy is an advanced and sophisticated art, application of this to acoustic microscopy is still in its infancy. The mechanisms of contrast for specimens whose acoustic properties are close to those of water are quite different from those for solids; surface smoothness is rather less important, the interference between reflections from the top and bottom of the specimen can dominate, and it is even possible to image living cells (Hilderbrand and Rugar, 1984) .

The prepared specimen should be finally cleaned with a light solvent such as ethanol, and it may now be mounted in the microscope and approximately levelled by eye. When the water droplet is later added for acoustic coupling, if it is not heavy enough surface tension may lift it up. To avoid this, light specimens should be fastened down with spring clips; double-sided sticky tape should not be used because of the variability in z that it would introduce. If the microscope is fitted with facilities for light microscopy the specimen may now be examined optically, and the approximate area to be imaged may be found. However it is not worth spending too much time positioning the specimen exactly, since it will move during the levelling process. For the same reason it is preferable to take any optical photographs after the acoustic micrographs have been obtained. If the optical and acoustic stages are parfocal, and the focusing adjustment has an upper stop fitted, this should now be set in order to reduce the likelihood of crashing the acoustic lens into the specimen. The specimen should now be lowered, the optical objective removed and the acoustic lens brought into position. The specimen stage should have been previously set for the correct temperature, 60–70° C for the water itself is usually optimum. Distilled water should be used, since although dissolved salts reduce the attenuation slightly they tend to be left behind as the water evaporates, leaving the specimen looking like the shores of the Dead Sea. However for ferrous specimens it is useful to add a few p.p.m. of sodium molybdate to the water to inhibit corrosion. The chief enemy at this stage is the formation of bubbles, this is exacerbated by the reduced solubility of gases in water as it is heated. Therefore freshly distilled water should be used, and the tendency to form bubbles will be reduced if it can be preheated prior to use.

A drop of water should now be applied to the lens. Up to this stage the small pulse between pulses two and four in Fig. 1.3 should not be seen. If the amplifiers are not too badly saturated it is possible that a reduction in the height of pulses two and four occurred as the drop of water was applied; this is because of the reduced reflection at the lens surface (from almost 100 per cent) when the water came into contact, and serves as a useful check that the water has indeed wetted it. Now the specimen must be brought up to the acoustic lens. The chief danger to avoid here is crashing the specimen into the lens. If this should happen the damage will be less if the lens is not vibrating; for this reason both scans should be off at this time. The specimen may be brought fairly close (within say 0.5 mm) to the lens by eye, but the difficulty then is to know what to watch. If a stop has been set on the coarse focus control, then the echo c.r.o. (c.f. Fig. 1.3) should be observed, with the hope of seeing the third echo corresponding to a reflection from the specimen. If no specimen echo has been found when the coarse focus has reached its end stop, the fine focus control may be used to raise the specimen a little further, while the lens is watched carefully for contact with the specimen. If by now an echo has been found, well and good; the problem is what to do if no echo has been obtained. Provided that echoes two and four of Fig. 1.3 can be seen then there can be no serious system fault, such as a bad connection to the lens or more likely wrong timing of the PIN switches (in which case consult the handbook of the microscope). The most likely fault is a bubble somewhere on the surface either of the lens or of the specimen. If no echo is obtained at all as the specimen is raised there is probably a bubble on the lens. In this case the specimen should be lowered and the lens blown dry with a low-pressure gas jet. A fresh drop of water should be applied and the process repeated until a specimen echo is found. The symptom of a bubble on the specimen is that an echo appears before the specimen is raised as far as it should be, the echo is erratic in behaviour (e.g. the height at which it is in focus is not constant), and finally when the fast scan is switched on a signal is seen only in the centre of the line scan. Bubbles on the specimen are generally less troublesome than bubbles on the lens; they are removed in a similar way, by lowering the specimen, drying it with a stream of gas, and applying some more water to the lens.

Once a specimen echo has been found, a feel for $V(z)$ may be obtained by raising and lowering the specimen (taking care not to drive it into the lens), and noting how the height of the specimen echo oscillates; this may be compared with Fig. 2.1. Indeed the microscope may have a facility for displaying $V(z)$ in real time. If so this can now be used to gain an idea of how the contrast may be expected to depend on defocus, and the specimen may be translated a little in the X and Y directions (having noted the settings that were chosen optically so that these may be returned to) to see how this changes across it. The task of levelling the specimen may now begin. It is best to start with the fast scan direction, so the fast scan should now be switched on, care being taken that the specimen is not so badly tilted, nor the scan so large, that there is danger of the lens scraping the specimen. A line scan should be seen that corresponds to details of the specimen superimposed on $V(z)$. This is because any tilt causes z to vary linearly across the specimen, so that in effect $V(z)$ is plotted by the line scan. The aim is to eliminate this. If the specimen is raised or lowered a little, the $V(z)$ component of the line scan will be seen to move from side to side while the specimen features will move up and down (corresponding to changes in intensity). A little geometrical thought will show that when the specimen is moved up $V(z)$ will move towards the side that is lower. The micrometer that controls tilt in the fast scan direction should now be turned so as to reduce the tilt that has been found. Initially a change of tilt of about 1 per cent should be tried; if the micrometer has a lever arm of 10 mm this corresponds to 10 divisions of a standard micrometer, or 0.1 mm. Great care should be taken while doing this. Unless the specimen happens to be positioned with the axes of the gimbals actually intersecting the lens axis, altering the tilt will alter z, so that either the lens could come into contact with the specimen (remember that the scan is now on) or the specimen could move a long way from the lens. In either case the signal could be lost, and finding it again can be tedious, especially if one does not know which way to alter the focus. It is therefore best to change the tilt with one hand on the fine focus control and one eye on the line scan. Once the tilt has been adjusted by the chosen amount the focus may be varied a little again to discover, from the direction of the movement of $V(z)$, whether the tilt is still in the same direction or whether the adjustment has overshot. In this way the tilt may be iteratively adjusted until there is no movement from side to side of the line scan as the fine focus is varied, and it simply moves up and down with the same defocus throughout its length. How accurately the tilt must be adjusted to achieve this depends on the nature of $V(z)$ and also on the ratio of the length scanned to the wavelength; typically the tilt needs to be correct to better than 10^{-3} radian. If the tilt cannot be adjusted to make the line scan move up and down uniformly with changes of defocus then either the lens is not scanning at a uniform height, or the specimen stage is vibrating vertically (it may be possible to reduce these two causes by varying the fast scan frequency), or, which is most likely, the specimen is not flat. To try to level an undulating specimen is probably the most frustrating task that can be devised for an acoustic microscopist, and would satisfy the 'object all sublime' of any Mikado!

The specimen must now be levelled in the other direction. Some microscopes have a facility for fast scanning in the *Y* direction, and in that case levelling may be carried out as for the *X* direction. But if not a different procedure must be adopted. It is best to leave the fast scan on, so that tilt can be observed throughout its length, so again care must be taken not to allow the lens to scrape the specimen. It is now vital to know where the specimen is on the $V(z)$ curve; the best place is on the positive z side of the peak, defocused away from the lens until the signal is about half its full strength, as here $V(z)$ has a steep slope and is fairly independent of specimen properties. The lens should then be driven in the *Y* direction, by either electrical or mechanical control depending on the facilities available, and the initial behaviour of the line scan noted. If the signal increases then the side of the specimen towards which the lens is moving is higher than the other side, if the signal decreases it is lower. The specimen tilt should be correspondingly adjusted, bearing in mind all the points mentioned under the adjustment of the X scan, and the iteration continued until a satisfactory constant mean level of the line scan is obtained when the lens is scanned over the whole of the area to be imaged.

Once the specimen has been levelled an image can be obtained. For this the optimum value of z must be chosen. A good idea of the dependence of the contrast on defocus can be obtained by watching the line scan as the defocus is varied. For this reason it is a good idea when imaging a specimen that has line features in one predominant direction to orient it with the features perpendicular to the fast scan. The changes in level of the line scan correspond to differences in brightness that will appear on the acoustic image taken with that value of z. Generally it is desirable, at least initially, to maximize the range of intensities on an image. This is achieved by altering z until the greatest variations in level are seen on the line scan; often this is found to be defocusing the specimen towards the lens until the mean level is about half its maximum (i.e. at focus). Having found this a scan can be made, and the image observed. The brightness and contrast levels of the monitor may need adjusting: in varying these it may be helpful to think of the displayed intensity ζ of a monitor as depending on the input signal ξ according to the transfer function $\zeta = b + c\,\xi$, b then corresponds to the 'brightness' control and c to the 'contrast' control (unfortunately some monitors spoil this by omitting black level clamping, in which case the transfer function becomes $\zeta = b + c(\xi - \frac{1}{2}\xi_0)$, where ξ_0 is the full range of ξ; this means that the screen does not go dark for zero input signal).

It is worth pausing at this point to appreciate the achievement of obtaining an image of optical microscope resolution using acoustic waves. From here on improving the image quality depends on knowing what one is looking for in the image, and understanding through the relevant contrast theory how the image may be optimized to reveal what is of most interest. The most useful parameter to vary is the defocus z (for example, it may sometimes be helpful to move to the second maximum of $V(z)$, though usually the best images are obtained somewhere on the slope of the principal maximum of $V(z)$ with negative z of minus two or three wavelengths), and it may be worth recording micrographs at different values of z (for example different layers of an i.c. may be revealed at different values of z). It may be possible to vary the equivalent of brightness and contrast on the input to the framestore; details of what is possible will vary from one microscope to another, and will quickly become familiar with experience. It is of course also necessary to choose the magnification which best reveals the features of interest; generally little is gained by using a magnification such that the area viewed is less than a hundred wavelengths across, and more useful micrographs may often be obtained at much lower magnification than this.

Once the possibilities of varying z and other parameters have been exhausted a final image should be obtained and recorded photographically. A record of each micrograph should include the photograph number, details of the specimen, the magnification, and the defocus $-z/\mu\text{m}$. It is not generally possible to read the absolute value of z from any scale, so this must be measured relative to focus (which is approximately the position of maximum signal). If an optical microscope is fitted, an optical image of the same area can now be made. The specimen should be lowered and both lens and specimen immediately dried with a jet of dry gas. The optical objective corresponding to the magnification that was used acoustically should be positioned, and the specimen brought into focus. If it is found to be dirty as a result of drying marks from the water these can be removed either by cleaning with a cotton bud soaked in ethanol, which will dry very rapidly because the specimen is hot, or, if they are deposits of the corrosion inhibitor (which is not soluble in ethanol), with more water dried off quickly with a cotton bud. Once the specimen is clean the optical image may be compared with the acoustic image, which should still be displayed by the framestore. How closely the area seen optically corresponds to that imaged acoustically depends on how accurately the two modes are aligned. Unless they are very badly out of adjustment, even at the highest magnification there should be some overlap, so that at least part of the area imaged acoustically should be visible. Some corresponding features should be found on the two images (remembering to allow for possible inversion if the optical image is viewed directly through the microscope), and the specimen may then be moved to align the two images exactly. An optical micrograph may then be recorded, and then either another area on the same specimen examined or the specimen removed from the microscope.

From time to time it may be necessary to clean the lens of the microscope. If it has become very dirty it should be removed and examined in a microscope or stereo microscope at about 100 × magnification. However in many cases this is not necessary, and it may be cleaned *in situ*, with its performance compared with that obtained on a day when it is known to have been giving good results (it is worthwhile recording these quantitatively when the instrument is new). To clean a lens with no anti-reflection coating, a cocktail stick may be sharpened to a fine point with a knife or scalpel. This point must then be inserted into the lens cavity (this can be found with a little practice) and then twiddled about its axis to loosen any dirt. If the lens has an acoustic matching layer it may not be as hard as sapphire, so this should be done gently, in order to avoid unnecessary wear. The lens should then be cleaned with a cotton bud soaked in a suitable solvent, twiddled this time about an axis perpendicular to the axis of the lens, and immediately dried with the other end of the cotton bud. The choice of solvent is important. It must not dissolve the material of the lens, particularly any epoxy that may have been used to mount and seal it. Ethanol is safe but not very powerful, trichloroethylene has been used very satisfactorily but dissolves certain polymers, acetone has been known to be fatal for a lens. In case of doubt it would be well to check with the lens manufacturer for his recommendation. In particular, if the lens has an acoustic anti-reflection coating of chalcogenide glass, neither solvents nor any rubbing should be used as both will cause permanent damage. The best practice of all is to dry the lens with pure nitrogen gas after every use; in this way it should never become dirty or need cleaning.

If the lens is removed for cleaning or for any other reason it is most unlikely that it will be replaced in exactly the same way, and even if it is the r.f. lead may be clamped with a slightly different orientation, and the alignment with the optical microscope may then be spoilt. To readjust this an object should be inserted with a feature that can easily be identified and located relative to any other point on the object. A circular t.e.m. grid is ideal for this purpose. It should be positioned so that it is concentric with the optical field of view at the highest magnification. It should then be examined acoustically. If a satisfactory image cannot be obtained without levelling it must be levelled and then again moved to the centre of the optical field of view, and then the acoustic lens repositioned. If the grid is thicker than the acoustic focal length then if the acoustic lens is above a hole in the grid the lens may meet the grid before any echo is found. In this case it may be better to break the earlier rule and switch on the fast scan before bringing the lens up, but first lay in a few spare grids! Once an acoustic image of the grid has been obtained, initially at low magnification to make it easier to know where one is, the offsets on the two scans must be adjusted until the acoustic image has the centre of the grid in its centre.

The other two checks that may be necessary on a regular basis are concerned with magnification and frequency. To check the magnification a t.e.m. grid may be inserted as a specimen and imaged. If the grid is, say, a British 400 mesh, then each square has side $\frac{1}{400}$ in, or 63.5 μm. Thus the magnification can be calibrated, and, most important, adjusted so that the magnification is the same in both X and Y directions. At the same time the frequency, amplitude and phase of the scans may be adjusted where this is possible and appropriate, to optimize the linearity of the image. To optimize the r.f. frequency, a $V(z)$ curve must be displayed. Some microscopes have a special facility for this, but if this is not available the easiest way to display $V(z)$ in real time is to insert a flat glass microscope slide as an object, and to let it be deliberately tilted in the X direction. The fast (X) scan should now be switched on, and because the glass surface is reasonably uniform, the variation of z because of the tilt will cause $V(z)$ to appear on the line scan display. The amount of tilt and the defocus should be adjusted until a good $V(z)$ is obtained (i.e. at least one minumum for z positive and three or four minima for negative z). The r.f. frequency should now be adjusted to give an optimal signal.

If fine structure is visible on $V(z)$ there are three possible causes. The first may be that the r.f. oscillator is not being properly isolated from the receiving amplifier. Assuming there is not a design fault in the r.f. system (e.g. insufficient isolation in the PIN switches) the only way to try to improve this is to vary the timing of the control pulses to the PIN switches. Secondly there may be acoustic reverberations in the lens from the previous pulse. If this is so it can be cured by reducing the pulse repetition frequency (p.r.f.), if there is an adjustment for this. But finally, there may be acoustic reverberations from the current transmitted pulse interfering with the echo from the specimen. It is part of good lens design to try to avoid these, but if they are present the only way to reduce them is to change the r.f. frequency. Thus in adjusting the r.f. frequency the first priority is to maximize the signal, and the second is to minimize any fine structure on $V(z)$. In maximizing the signal the effect of attenuation in the coupling fluid should be remembered; thus if two frequencies are found that give signals of similar strength, the higher frequency probably has more efficient transduction, and in any case is generally preferable because it gives better resolution.

3

Theory of scanning acoustic microscopy

A small amount of experience in using a reflection scanning acoustic microscope (s.a.m.) quickly reveals that the contrast is strongly dependent on the defocus. This was mentioned in the last chapter when it was suggested that once a line scan had been obtained the focus should be varied to find the maximum differences along the line scan, and as this is done the mean level of the line scan also changes. If images are taken at greatly differing values of the defocus, z, contrast reversals may even occur, so that the relative brightness of two regions on the image change over. Interpretation of contrast in the s.a.m. is therefore not a simple matter. One cannot simply say that a brighter area corresponds to a higher (or lower) density, or to a greater (or smaller) elastic modulus. Rather, interpretation must be based on the variation of the signal intensity (V, the voltage generated at the transducer) with defocus (z). This chaper is therefore concerned with the theory of $V(z)$ when the imaging scans are turned off, and the specimen is scanned along the axis of the lens. This may then be used in the interpretation of images scanned at a given value of z.

A high resolution acoustic lens with water as the coupling fluid cannot generally be used to focus below the surface of a solid. The velocity of longitudinal acoustic waves in most solid materials is typically $6000\,\mathrm{m\,s^{-1}}$. The velocity in water is approximately $1500\,\mathrm{m\,s^{-1}}$, giving a refractive index of 0.25 and a critical angle of about 15°. Waves incident at this angle excite longitudinal waves parallel to the surface, and above it do not excite any longitudinal waves in the solid at all. Thus small-aperture lenses would have to be used, which would make the resolution poorer. But now consider the pulse length. If the shortest pulse that can be made is 20 ns long, then for the echo from a subsurface feature to be time-resolved the feature must be at least $60\,\mu\mathrm{m}$ below the surface, otherwise the echo will be swamped by the surface echo, which is much larger because of the transmission losses in passing each way through the surface. However, because of the real and apparent depth problem even the paraxial rays appear to come from four times this depth, while rays at larger angles to the axis in the solid appear to come from even deeper. Thus the lens would need to have a focal length, measured from its rim, greater than $240\,\mu\mathrm{m}$. The highest frequency that could operate with a lens of that size is about 500 MHz, and because of the small numerical aperture (or equivalently because of the longer wavelength in the solid) the resolution of this would be worse than $10\,\mu\mathrm{m}$. For polycrystalline materials the frequency must be lower still because of grain scattering, and the highest frequencies that are normally used for water-coupled subsurface imaging are below 100 MHz, though an example of subsurface imaging in polycarbonate, which has acoustic properties closer to those of water than most solids, will be mentioned in the Chapter 4. There has been one attempt to use liquid gallium to couple to shear waves for subsurface imaging in a high velocity material (Jipson 1979), but the experiment has never been repeated; subsurface imaging in semiconductors using mercury as the coupling fluid promises to be more fruitful (Attal 1983). While such microscopy is very important, the imaging theory is relatively straightforward and will not be discussed here. In high resolution microscopy with wide angle lenses the contrast derives from the reflection at the surface of the specimen, and the theory of this will now be considered (Atalar 1978; Quate, Atalar, and Wickramasinghe 1979).

The situation may be described as follows (Sheppard and Wilson 1981). A plane wave of unit amplitude is generated by the transducer, and is refracted by a lens of pupil function for waves travelling in this direction of $P_1(\theta)$, where θ is the angle subtended with the axis of the lens at the focus and the pupil function describes the amplitude of sound transmitted through the lens. The amplitude is then

$$U_1(\theta) = \cos^{1/2}\theta \cdot P_1(\theta). \tag{3.1}$$

The $\cos^{1/2}\theta$ term is put in only for the completeness, it cancels out later. The wave is reflected at the focus by an object with a reflectance function $R(\theta)$, and then has amplitude

$$U_2(\theta) = \cos^{1/2}\theta \cdot P_1(\theta) \cdot R(\theta). \tag{3.2}$$

This reflected wave is refracted by the lens again and arrives at the transducer; in this direction the lens-transducer pupil function is $P_2(\theta)$. The acoustic field at the transducer is then

$$\begin{aligned} U_3(\theta) &= P_1(\theta) \cdot R(\theta) \cdot P_2(\theta) \\ &= P^*(\theta) \cdot R(\theta) \end{aligned} \tag{3.3}$$

where $P^*(\theta) = P_1(\theta) \cdot P_2(\theta)$, and is a composite pupil function for the lens and transducer together. Lens aberrations can be allowed for by letting P^* be complex. The signal at the transducer is obtained by summing this over the area of the transducer, with radial co-ordinate r, to give

$$V(0) = \int_0^\infty P^*(\theta) \cdot R(\theta) \cdot 2\pi r \cdot \mathrm{d}r. \tag{3.4}$$

Substituting $f \cdot \sin\theta = r$, $f \cdot \cos\theta \cdot \mathrm{d}\theta = \mathrm{d}r$, where f is the focal length, this may be written

$$V(0) = 2\pi f \int_0^{\pi/2} P^*(\theta) \cdot R(\theta) \cdot \sin\theta \cdot \cos\theta \cdot \mathrm{d}\theta. \tag{3.5}$$

If the reflecting surface is now moved towards the lens by a displacement $-\mathbf{z}$ from the focus, the phase of the wave incident at a given point on the surface will advance by $\mathbf{k} \cdot \mathbf{z}$ (where $\mathbf{k}$ is the wavevector in the fluid), and waves returning to the lens will advance their phase by twice this, so that the signal at the transducer is now

$$V(z) = 2\pi f \int_0^{\pi/2} P^*(\theta)\cdot R(\theta)\cdot e^{-i2kz\cdot\cos\theta}\cdot\sin\theta\cdot\cos\theta\cdot d\theta. \tag{3.6}$$

In practice the integration need not be carried as far as $\frac{1}{2}\pi$ because the pupil function will have cut off before that. For a given lens the pupil function will be constant, so that $V(z)$ will be determined by the reflectance function of the specimen, which in turn depends on the elastic properties.

The propagation of acoustic waves in a solid is more complicated than in a fluid. In a fluid sound travels as a longitudinal wave; it cannot have a polarization and at a given temperature and pressure in a particular medium there is a unique velocity. A solid is characterized by its ability to support tensor stresses rather than simply a scalar pressure. The elastic waves are then described by the Christoffel equations, which have three roots (Auld 1973). For an isotropic solid one of these corresponds to a longitudinal wave, and the other two are degenerate shear waves, with orthogonal polarizations and velocity typically half that of the longitudinal wave (exactly a half for Poisson's ratio $\sigma = \frac{1}{3}$). A third kind of acoustic wave can occur on the surface of an elastic solid. This is a Rayleigh wave, it is unique to acoustics because it depends on the ability of the medium to support both shear and longitudinal stresses; it is bound to the surface and its amplitude decays exponentially away from the surface. The velocity of a Rayleigh wave is typically 0.93 of the shear wave velocity, the exact proportion depending on Poisson's ratio for the material: a useful approximate relationship is $V_R/V_s \simeq (0.87 + 1.12\sigma)/(1 + \sigma)$. If the surface is in contact with fluid they should be called leaky Rayleigh waves, because they radiate (or 'leak') energy into the fluid, though for brevity the term leaky will be omitted here. The waves generated in the fluid by the Rayleigh wave propagate at the Rayleigh angle, which is the angle that satisfies Snell's law for this situation. Because Rayleigh waves are bound to the surface, they can interact with the fluid over appreciable distances, and for this reason the acoustic coupling to waves in the fluid can be much stronger than it is for bulk waves in a solid. Rayleigh waves therefore play a dominant role in the contrast in acoustic microscopy of solids. A table of elastic wave velocities for a number of solids will be found at the end of this chapter (Table 3.1).

The reflectance function for waves in a fluid incident on the surface of an isotropic solid may be derived in a form suitable for calculation on a computer as follows. The variables are defined with reference to Fig. 3.1: c and θ are the velocity and angle of incidence of the waves in the fluid, whose density is ρ; c_1, θ_1 and β_1, γ_1 are the velocities and angle of refraction of the longitudinal and shear waves in the solid, whose density is ρ_1 (the notation of Brekhovskikh (1980) has been retained to facilitate reference to that book). Snell's law relates θ_1 and γ_1 to θ:

$$\frac{\sin\theta}{c} = \frac{\sin\theta_1}{c_1} = \frac{\sin\gamma_1}{b_1}. \tag{3.7}$$

Impedances are then defined by

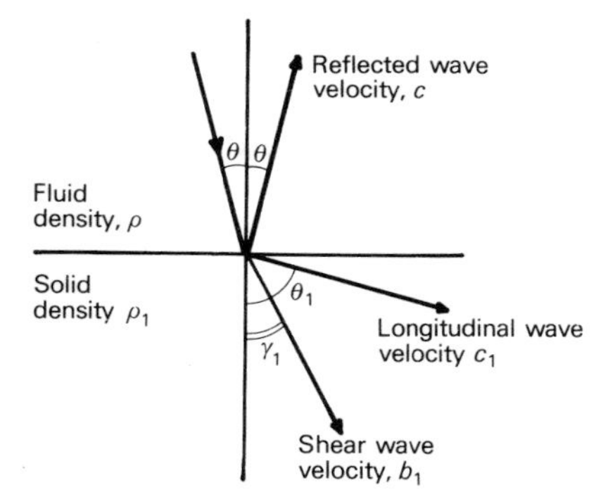

Fig. 3.1. The reflected and transmitted rays at a fluid–solid interface, described by eqns. (3.7–3.10).

$$Z = \frac{\rho c}{\cos\theta}, \quad Z_\ell = \frac{\rho_1 c_1}{\cos\theta_1}, \quad Z_t = \frac{\rho_1 b_1}{\cos\gamma_1} \tag{3.8}$$

$$Z_{tot} = Z_\ell \cos^2 2\gamma_1 + Z_t \sin^2 2\gamma_1. \tag{3.9}$$

Then the reflectance function is

$$R(\theta) = \frac{Z_{tot} - Z}{Z_{tot} + Z}. \tag{3.10}$$

The reflectance function for acoustic waves incident from water onto glass is shown in Fig. 3.2. $R(\theta)$ is a complex function, so the modulus is referred to the left ordinate, and the phase to the right ordinate. Starting from zero angle of incidence, the modulus of the reflectance function experiences its first major fluctuations at around 15°, this corresponds to the critical angle for longitudinal wave excitation in the glass. Above the shear wave critical angle no energy can be propagated into the solid, so the modulus of the reflectance function must be unity. The phase of the reflectance function also experiences small fluctuations around the longitudinal critical angle, but its most dramatic behaviour occurs a few degrees beyond the shear wave critical angle. This is centred around the angle at which waves in the liquid can couple into Rayleigh waves in the surface of the solid, and as can be seen the phase changes by almost 2π over a fairly small change in the angle of incidence. This phase change dominates the behaviour of $V(z)$; to see why, it helps to turn the expression for $V(z)$ into a Fourier transform. This is achieved by the substitution

$$t = 2k\cdot\cos\theta$$

$$dt = -2k\cdot\sin\theta\cdot d\theta \tag{3.11}$$

then

$$V(z) = 2\pi f \int_0^{2k} P'(t/k)\cdot R'(t/k)\cdot e^{-itz}\cdot\frac{t}{4k}\cdot dt$$

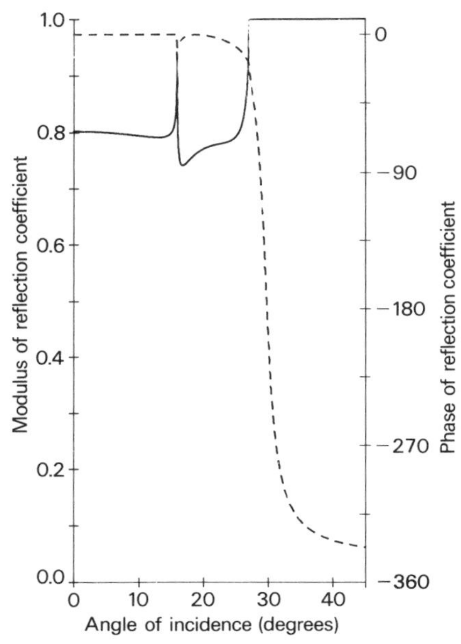

Fig. 3.2. The reflectance function $R(\theta)$ for waves in water incident on a glass surface.

$$= \frac{\pi f}{2k} \int_{-\infty}^{\infty} P'(t/k) \cdot R'(t/k) \cdot e^{-itz} \cdot t \cdot dt \tag{3.12}$$

where P' and R' are suitably defined for the new argument, and the integral may be taken over infinite limits, provided $R' = 0$ unless $0 \leqslant t/k \leqslant 2$. This expression can be recognized as a Fourier transform, with t and z as the tranform pair. Consider first the case when $R = 1$ for all angles of incidence (Atalar 1979). This would correspond to reflection from a rigid surface or from a boundary with a vacuum with a change of sign. If the pupil function were unity up to some angle α, and zero thereafter (this is sometimes called a top hat function, because of its appearance when visualised in three dimensions), then $V(z)$ is simply the transform of this, which for small α is a sinc^2 function (sinc $\phi \equiv \sin \phi/\phi$, this is the diffraction pattern from a single slit). Of course α is not small, but nevertheless this behaviour is approximately followed in the microscope, especially for $z \geqslant 0$, and it accounts for the central maximum in $V(z)$ near focus.

If now we insert the reflectance function of a real material, such as that shown in Fig. 3.2, the most prominent feature is the phase change of almost 2π around the Rayleigh angle. In the new variables this occurs at $t = 2k \cdot \cos \theta_R$, and there is also a discontinuity corresponding to $\theta = 0$, which in the new variables is at $t = 2k$. In the Fourier transform this will give rise to oscillations of periodicity

$$\Delta z = \frac{2\pi}{2k - 2k \cdot \cos \theta_R} \tag{3.13}$$

where k is the wavenumber in the fluid.

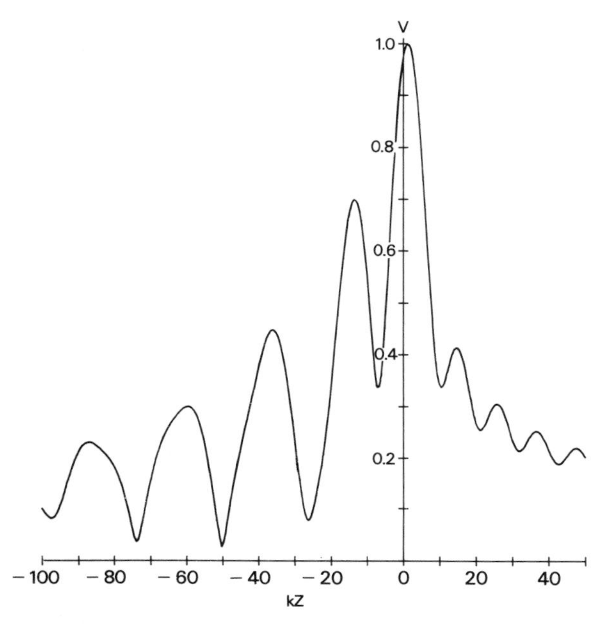

Fig. 3.3 $V(z)$ for glass, with water as the coupling fluid.

$V(z)$ for the material of Fig. 3.2 is computed in Fig. 3.3. This enables the main features of a theoretical $V(z)$ function to be seen. First there is the strong central maximum which is determined by the pupil function of the lens. It may be noted that this does not occur exactly at $z = 0$, but the difference is usually slight. Secondly the curve for positive z looks similar to a sinc^2 function, the value for the first minimum in a sinc^2 function for this value of α is at $kz = 11$. On the negative z side there are the strong oscillations, the periodicity for these that would be predicted by the above formula is $\Delta kz = 23$. Experimental $V(z)$ curves are often plotted with a logarithmic V-axis, this can make the minima appear as very pronounced nulls.

In practice it is very difficult to characterize the pupil function, expecially for a high resolution lens. This is the main reason for the discrepancy between Fig. 3.3 and experimental results such as that in Fig. 2.1. The pupil function used in the calculation of Fig. 3.3 was a top hat of semi-angle $\alpha = \frac{1}{4}\pi$. Experimental curves are obtained using a square law detector, for this reason the theoretical curves presented here have been squared; they should perhaps be called $V^2(z)$ curves, though the convention is to refer to them simply as $V(z)$. The periodicity of the oscillations, however, is not dependent on the pupil function, and there is good correlation between eqn (3.13) and experiment. This is illustrated in Fig. 3.4, which presents the results of the measurement of the oscillation periodicity for a number of materials plotted against their calculated Rayleigh angle (Weglein 1979*a*).

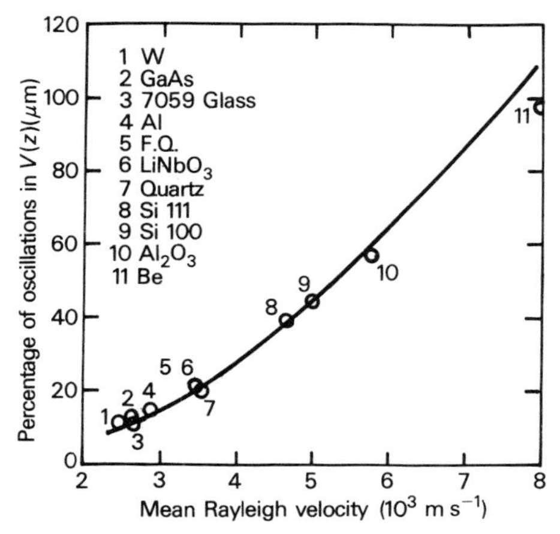

Fig. 3.4. The experimental periodicity of the oscillations in $V(z)$ plotted against elastic wave velocity for a number of materials (350 MHz; Weglein 1979*a*). The solid curve is equation (3.13/15); this is a slight improvement on the curve originally published in 1979.

The periodicity scales with wavelength, and since the radius of a lens is made proportional to the square of the wavelength, the number of oscillations that can be measured before the specimen meets the lens decreases at higher frequencies. The measurements of Fig. 3.4 were made at 400 MHz, at which about five oscillations could be measured. Also shown is the theoretical curve for the oscillation spacing as a function of the Rayleigh angle, and the agreement is remarkably good. Because Rayleigh wave velocity is such an important parameter in understanding contrast it is listed for some of the materials in the table at the end of this chapter. It is also invaluable to have a computer program available (this requires no more than a fast Fourier transform routine or numerical integration) to calculate $R(\theta)$ and $V(z)$ for any specimen being studied. $V(z)$ for more complicated situations will be discussed in Part II. The Fourier transform may be inverted to yield $R(\theta)$ if the amplitude and phase of $V(z)$ are measured and $P(\theta)$ has once been calibrated (Hildebrand, Liang, and Bennett 1983). This may eventually enable the acoustic microscope to be operated as a quantitative microprobe.

The Fourier analysis emphasizes the role of two waves, incident at the Rayleigh angle and normally. This justifies the use of a simple ray model, illustrated in Fig. 3.5, to predict the periodicity of the oscillations in $V(z)$ (Parmon and Bertoni 1979). Compared with focus the normal ray undergoes a change in path length of z before being reflected, and a further amount z on the return journey, i.e. a total change in path length of $2z$. The ray incident at the Rayleigh angle has its path in water shortened by $z \cdot \sec \theta_R$, and then excites a Rayleigh wave on the specimen surface. This propagates along the surface leaking a wave back into the fluid, and the ray symmetrically placed with respect to the incident ray travels back to the lens with its path also shortened by $z \cdot \sec \theta_R$. The reason why this ray is the one that counts comes from consideration of stationary phase; because the transducer

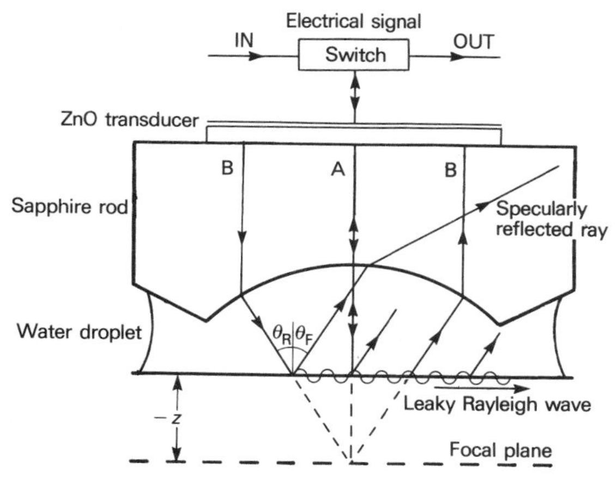

Fig. 3.5. Ray model for determining the periodicity in $V(z)$.

sums all waves incident on it, any wavefront arriving at other than normal incidence allows contributions from one part of the wavefront to cancel those from another part, so that the net signal is small. Only the ray symmetrical to the ray that excites Rayleigh waves arrives normally at the transducer, and therefore produces a signal. The path length travelled by the Rayleigh wave must now be added: this is $2z \cdot \tan \theta_R$, but because of the difference in velocities this is equivalent to an acoustic path in the fluid of $2z \cdot \tan \theta_R \cdot \sin \theta_R$, since by Snell's law $c_f/c_R = \sin \theta_R$. There is also a phase change of π associated with the excitation and re-radiation of the Rayleigh wave. The phase difference between the two rays (assuming that the lens eliminates all other phase differences) is

$$\begin{aligned} \Delta\phi &= (2z - 2z \cdot \sec \theta_R + 2z \cdot \tan \theta_R \cdot \sin \theta_R) \cdot k + \pi \\ &= 2z \cdot (1 - \sec \theta_R \cdot \{1 - \sin^2 \theta_R\}) \cdot k + \pi \\ &= 2kz \cdot (1 - \cos \theta_R) + \pi \, . \end{aligned} \tag{3.14}$$

This will give oscillations of periodicity

$$\Delta z = \frac{2\pi}{2k \cdot (1 - \cos \theta_R)} \tag{3.15}$$

in agreement with the result obtained using Fourier optics (3.13) (Atalar 1979; Wickramasinghe 1983). The methods of ray theory can also be used to give a good account of many other aspects of $V(z)$ (Bertoni 1984).

In conclusion, the contrast in the s.a.m. is generally described by the $V(z)$ formalism. The dominant feature in this is the behaviour of the reflectance function around the Rayleigh angle. Therefore if good contrast is to be obtained it is essential to use a lens of numerical aperture large enough to include the Rayleigh angle. Of course this may not apply to very soft biological specimens: if these are in the form of a thin layer mounted on a glass substrate, they may progressively quench the Rayleigh waves, and eventually interference between reflections from the top and bottom surfaces of the specimen may dominate. But for the imaging of solid materials the theory described here is very important.

Table 3.1. *Elastic wave velocities for various solids in the region of 20° C are given below. In the case of metals, factors such as texture, cold work, stress, hardening, temperature and aging can cause significant departures from the values given in the table. Properties of plastics vary considerably with molecular weight, with additives and with temperature. In view of this, many values are approximate and relate to materials of variable composition of materials is given as percentage by weight.* (*Kaye and Laby 1966*)

Material	Wave velocity, m s^{-1}			Material	Wave velocity, m s^{-1}		
	c_l longitudinal bulk waves	c_s shear waves	c_R Rayleigh waves		c_l longitudinal bulk waves	c_s shear waves	c_R Rayleigh waves
Aluminium	6374	3111	2906	Niobium	5068	2092	1970
Barium titanate ceramic	4000	–	–	Nylon	2680	–	–
Beryllium	12890	8880	–	Perspex	2700	1330	1242
Bone, human tibia	4000	1970	–	Platinum	3260	1730	–
Brass	4372	2100	1964	Polythene (polyethylene)	2000	–	–
Cadmium	2780	–	–	Polystyrene	2350	1120	1047
Cellulose acetate butyrate	2080	–	–	Polyvinyl chloride	2300	–	–
Chromium	6608	4005	3655	Polyvinyl chloride acetate	2250	–	–
Constantan	5177	2625	2445	Polyvinyl formal	2680	–	–
Copper	4759	2325	2171	Polyvinylidene chloride	2400	–	–
Duralumin	6398	3122	2917	Quartz (crystal) X-cut	5720	–	–
Ebonite	2500	–	–	Quartz fused	5970	3765	3410
Glass (crown)	5660	3420	3127	Silica (fused)	5968	3764	–
Glass (heavy flint)	5260	2960	2731	Silver	3704	1698	1592
Glass (pyrex)	5640	3280	–	Steel (mild)	5960	3235	2996
Gold (hard-drawn)	3240	1200	–	Steel (tool) hardened	5874	3179	2945
Granite	5400	–	–	Steel (stainless)	5980	3297	3049
Invar	4657	2658	2447	Tantalum	4159	2036	1902
(36 Ni, 63.8 Fe, 0.2 C)				Tin	3380	1594	1491
Iron (soft)	5957	3224	2986	Titanium	6130	3182	2958
Iron (cast)	4994	2809	2590	Tourmaline (crystal) Z-cut	7250	–	–
Lead	2160	700	–	Tungsten (annealed)	5221	2887	2668
Magnesium	5823	3163	2930	Tungsten (carbide)	6655	3984	3643
Manganese	4600	–	–	Tungsten (drawn)	5410	2640	–
Molybdenum	6475	3505	3248	Uranium	3370	1940	–
Monel metal	5350	2720	–	Vanadium	6023	2774	2600
Nickel (unmag. soft)	5608	2929	2722	Zinc (rolled)	4187	2421	2225
Nickel (unmag. hard)	5814	3078	2857	Zirconium	4650	2250	–

Part II: The Image

4

Elastic properties

The scattering of acoustic waves from abrupt changes in elastic properties is familiar from ultrasonic non-destructive testing: there a flaw inside a specimen causes scattering because the fluid in the flaw has a much lower acoustic impedance than the surrounding material. Similar scattering from subsurface features can be obtained in a low resolution s.a.m. (Nikoonahad 1984). In Fig. 4.1 an acoustic image is shown of the bond between a transistor and its heat sink. This is taken in reflection, and so the bright areas of the picture, corresponding to high acoustic intensity, show where the bond is poor. In these pictures the frequency used was 60 MHz, giving a resolution of about 0.1 mm, and the plane to be imaged was about 1 mm below the top surface. As in ultrasonic n.d.t., there is a dead zone below the specimen surface in which any echo from a

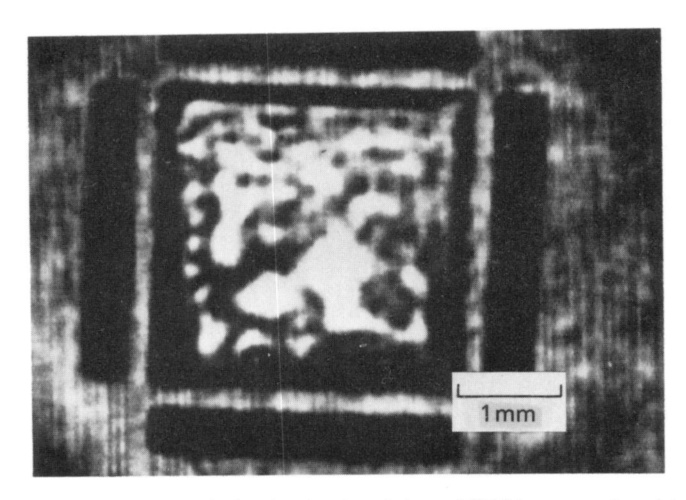

Fig. 4.1. An acoustic image of the header bond in a TIP33A power transistor (60 MHz, Nikoonahad, Yue, and Ash 1982).

defect would be swamped by the echo from the surface. For reasons discussed in the last chapter, this dead zone can never be made much less than about 60 μm deep, and because of the foreshortening of the focal length owing to the difference between real and apparent depth, the high resolution s.a.m. must normally be considered as a technique for looking at surface or thin films. The reflection from a discontinuity in impedance is given by

$$R = \frac{Z_2 - Z_1}{Z_2 + Z_1} \tag{4.1}$$

For a typical solid–water interface, Z_1 is about 30 Mrayl and $Z_2 = 1.5$ Mrayl, so that $R = 0.98$ and 96 per cent of the incident power is reflected (Chapter 3). Therefore the scattering from a subsurface flaw in a low resolution s.a.m. is strong, and no special theory is needed to account for it. The same formula may be applied to predict the change in reflection of normally incident waves from the surface of a specimen due to variations in elastic properties.

$$\Delta R = \frac{2Z_1}{(Z_2 + Z_1)^2} \cdot \Delta Z_2 \tag{4.2}$$

giving

$$\frac{\Delta R}{R} = \frac{2Z_1 Z_2}{Z_2^2 - Z_1^2} \cdot \frac{\Delta Z_2}{Z_2} \tag{4.3}$$

Now for waves in water being reflected from a solid, the impedance of the solid Z_2 is much greater than the impedance of the water Z_1, typically $Z_2/Z_1 \approx 15$. Therefore a change of say 30 per cent in the impedance of the solid would make a difference of only 2 per cent in the amplitude of the reflected wave; the coefficient of reflection is so nearly unity that its departure from unity is only weakly dependent on the impedance of the solid. If this were the whole story, the s.a.m. would be very insensitive to the elastic properties of solids. Fortunately, as it is defocused the microscope becomes sensitive to changes in the reflectance function around the Rayleigh angle. The theory for this effect was given in Chapter 3. The resolution is scarcely degraded by modest amounts of defocus (Smith, Wickramasinghe, Farnell, and Jen 1983).

Since the variation of contrast with defocus, $V(z)$, depends on the reflectance function $R(\theta)$, which in turn depends on the elastic properties, changes in elastic properties should affect the contrast. The most straightforward application of this is to image specimens that contain different materials. In Fig. 4.2 optical (a) and acoustic (b) images are presented of a specimen containing particles of ore that have been crushed after being mined. Here an example has been chosen that gives reasonable contrast optically; even so the acoustic contrast is so great that it is difficult to expose the finer contrast within both materials of the particle simultaneously. There are other mining products that give little or no contrast optically, and because the acoustic contrast arises from quite different properties it may be useful in assessing the degree of separation in these cases. Figure 4.3 shows another example. This is a glass-fibre-reinforced composite. It is a cross-ply laminate, and

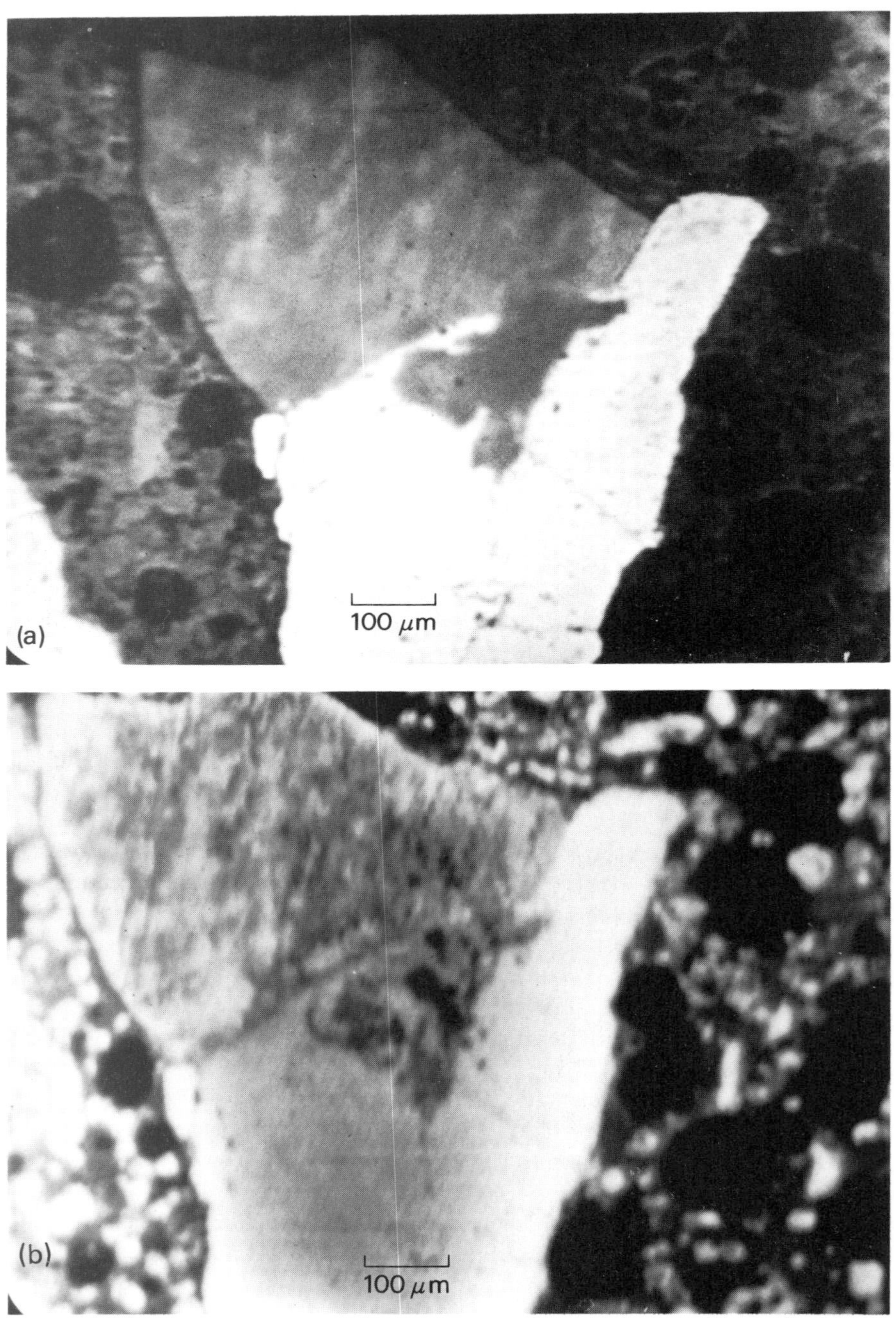

Fig. 4.2. A mining particle containing two kinds of mineral: (a) optical; (b) acoustic (0.73 GHz).

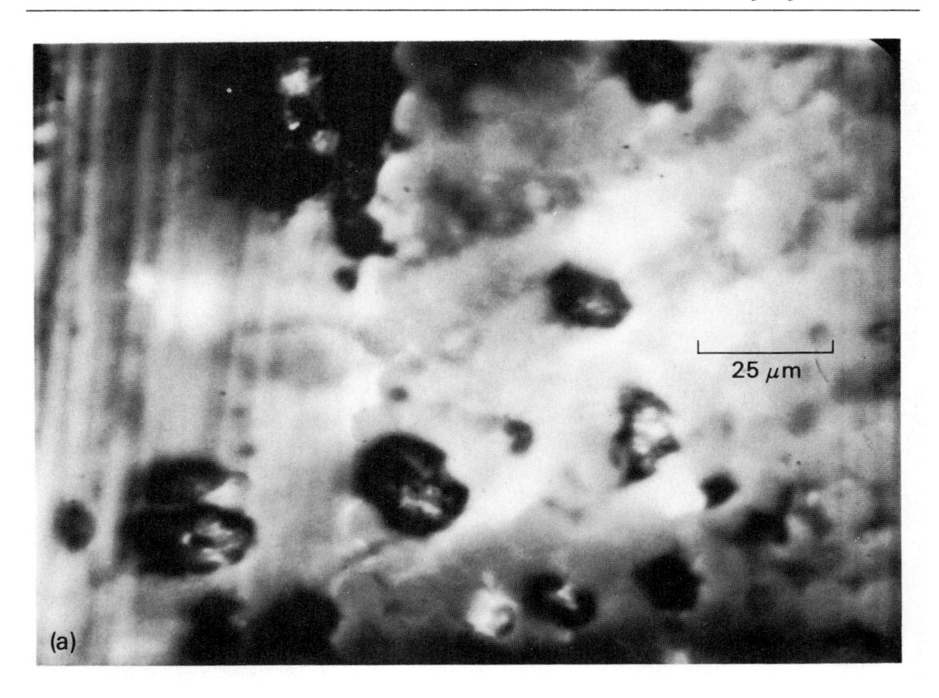
25 μm
(a)

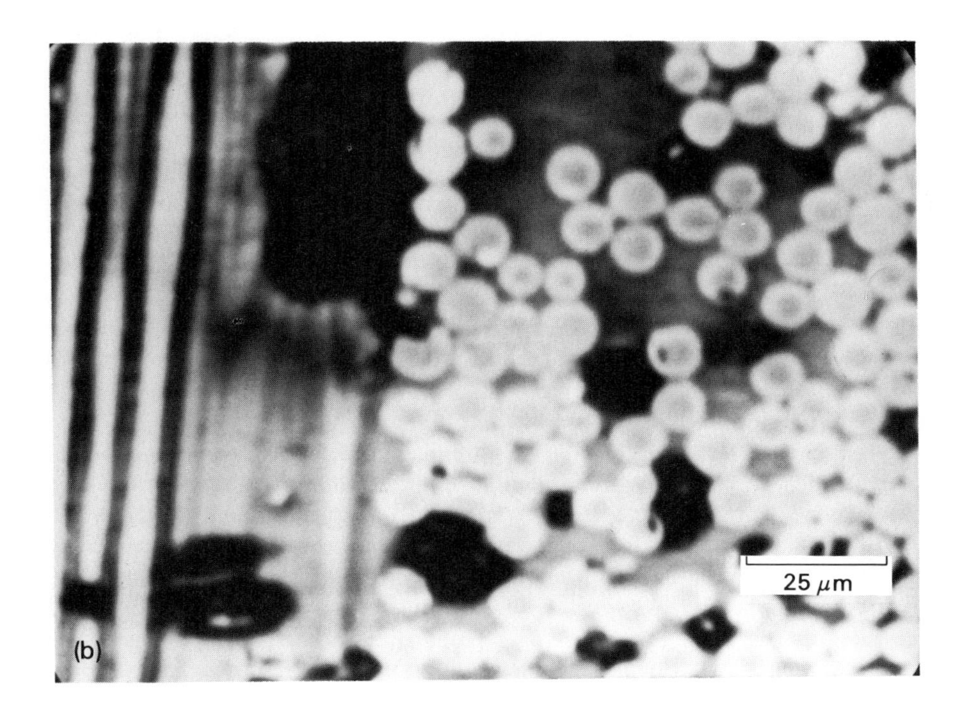
25 μm
(b)

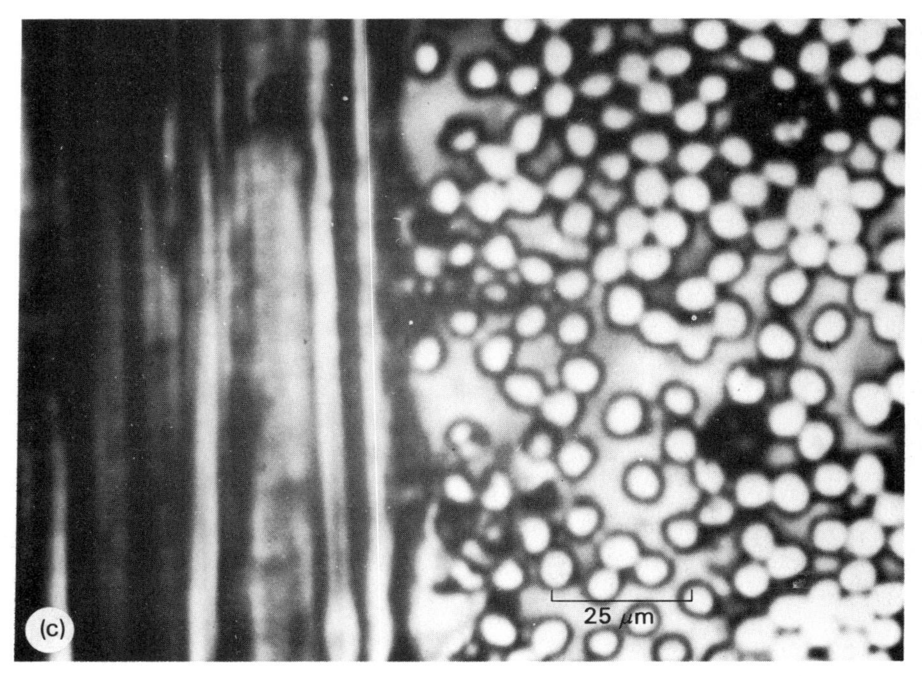

Fig. 4.3. A glass fibre reinforced epoxy cross-ply composite: (a) optical; (b) acoustic (0.73 GHz), (c) acoustic at greater defocus (this will be discussed in Chapter 7).

the optical (a) and acoustic (b, c) images show a region that includes two plies, parallel (left) and perpendicular (right) to the page. Both fibre and epoxy are transparent, so that little contrast is visible optically. They have quite different elastic properties, however, so that good contrast is found acoustically. Fascinating images have also been obtained of glass fibres embedded in a polycarbonate matrix (Hollis, Hammer, and Al-Jaroudi 1984). The velocity of longitudinal acoustic waves in the matrix is only 2000 m s^{-1}, so that in that case it was possible to image with high resolution below the surface, and subsurface imaging to a depth of 105 μm was achieved at 0.8 GHz. It must be emphasized that this is an exception to the general observation that the high resolution s.a.m. images the properties near the surface of a specimen.

It is important in any kind of reflection microscope to ask what depth of material is being sampled. The theory outlined in Chapter 3 suggested that Rayleigh waves play a vital role in the differences in contrast between different materials. An equivalent skin depth for Rayleigh waves can be defined; this is approximately equal to a Rayleigh wavelength, which at 0.73 MHz is 4 μm. An example illustrating this is presented in Fig. 4.4, which is of a specimen of superalloy that has so-called necklaces of voids. During the polishing in preparation for metallography, metal can be smeared over these, leaving subsurface voids. The light microscope samples only

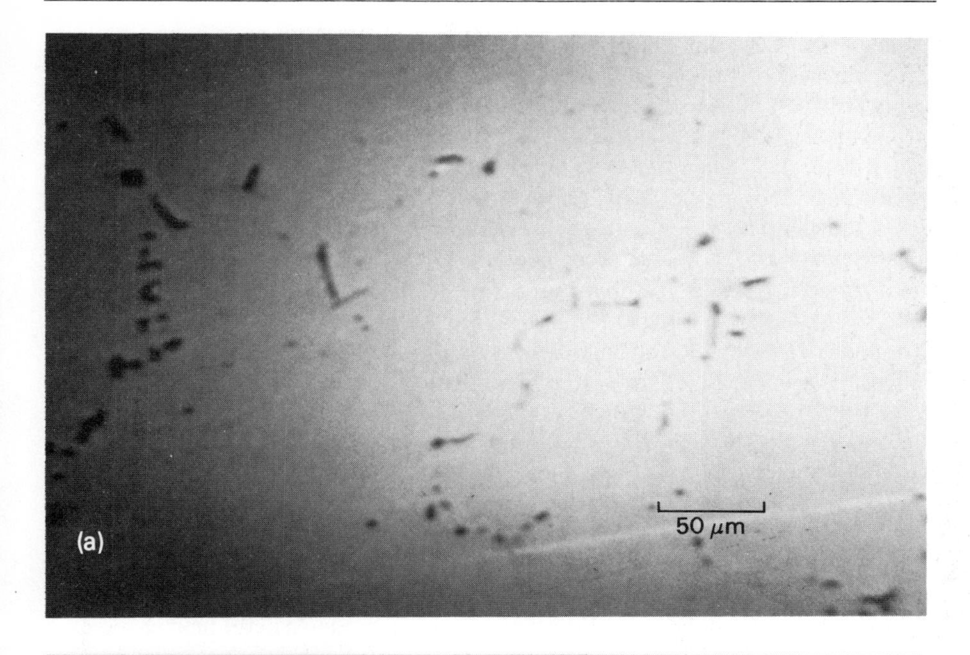

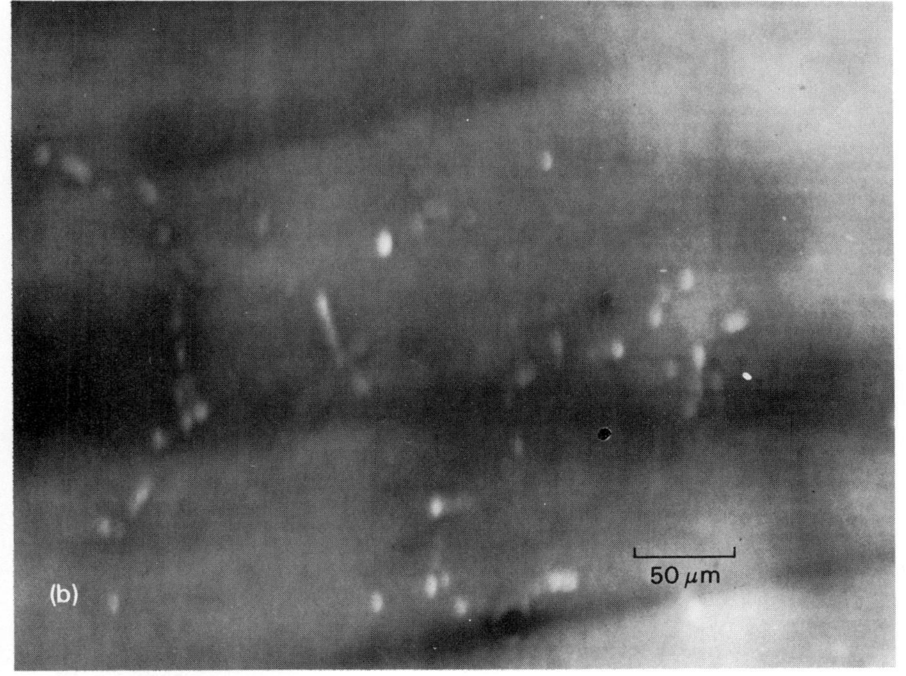

Fig. 4.4. A gas-turbine superalloy, with necklaces of voids: (a) optical; (b) acoustic (0.73 GHz).

the electromagnetic skin depth, which at optical frequencies is about 1 nm, so that if the thickness of the smeared-over layer is greater than this (which in practice it usually will be), the surface appears uniform. The acoustic microscope, however, is able to probe through the smeared layer. If there is a subsurface void, the Rayleigh wave will locally have the character of a Lamb wave in a plate, so that the wavelength will be longer here. This has the effect of broadening the central maximum in $V(z)$, so that the region with a subsurface void will appear as a bright spot in the acoustic image for all values of defocus up to the first minimum in $V(z)$ for the parent material. Some of the bright spots that can be seen in Fig. 4.4(b) do not correspond to any features visible in Fig. 4.4(a). Further evidence that the depth sampled corresponds to about a Rayleigh wavelength is shown in Fig. 4.5, which is an acoustic image of PLZT, a transparent ferroelectric ceramic (Yin, Ilett, and Briggs 1982). The surface of the ceramic was polished but not etched, and a strip of gold of thickness 0.1 μm was evaporated onto the surface. The acoustic image shows the grain structure, even under the gold. This suggests that the s.a.m. is sampling a layer considerably thicker than the gold, though not so thick that images of grains become hopelessly confused by grains further beneath the surface. The effect of the gold layer is to increase the average intensity of reflection; this may be partially understood in terms of conventional thin film interference arguments. The origin of the contrast from the grain structure will be considered in the next chapter.

As a final example in this chapter optical (a) and acoustic (b) images of human teeth enamel examined in cross-section are presented in Fig. 4.6. Enamel consists of crystals of hydroxyapatite in an organic matrix of collagen. When there is an

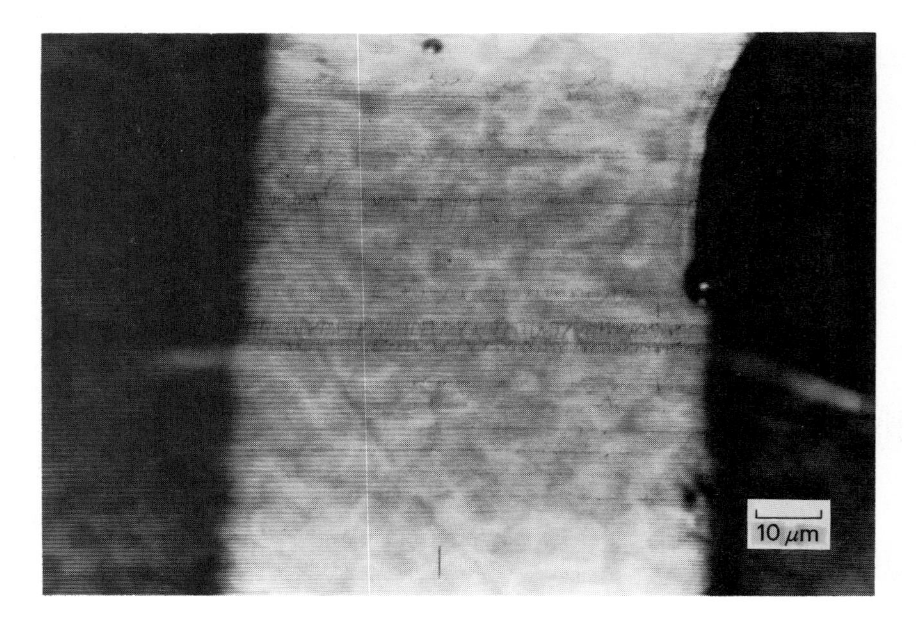

Fig. 4.5. PLZT ceramic, with a gold strip (0.73 GHz).

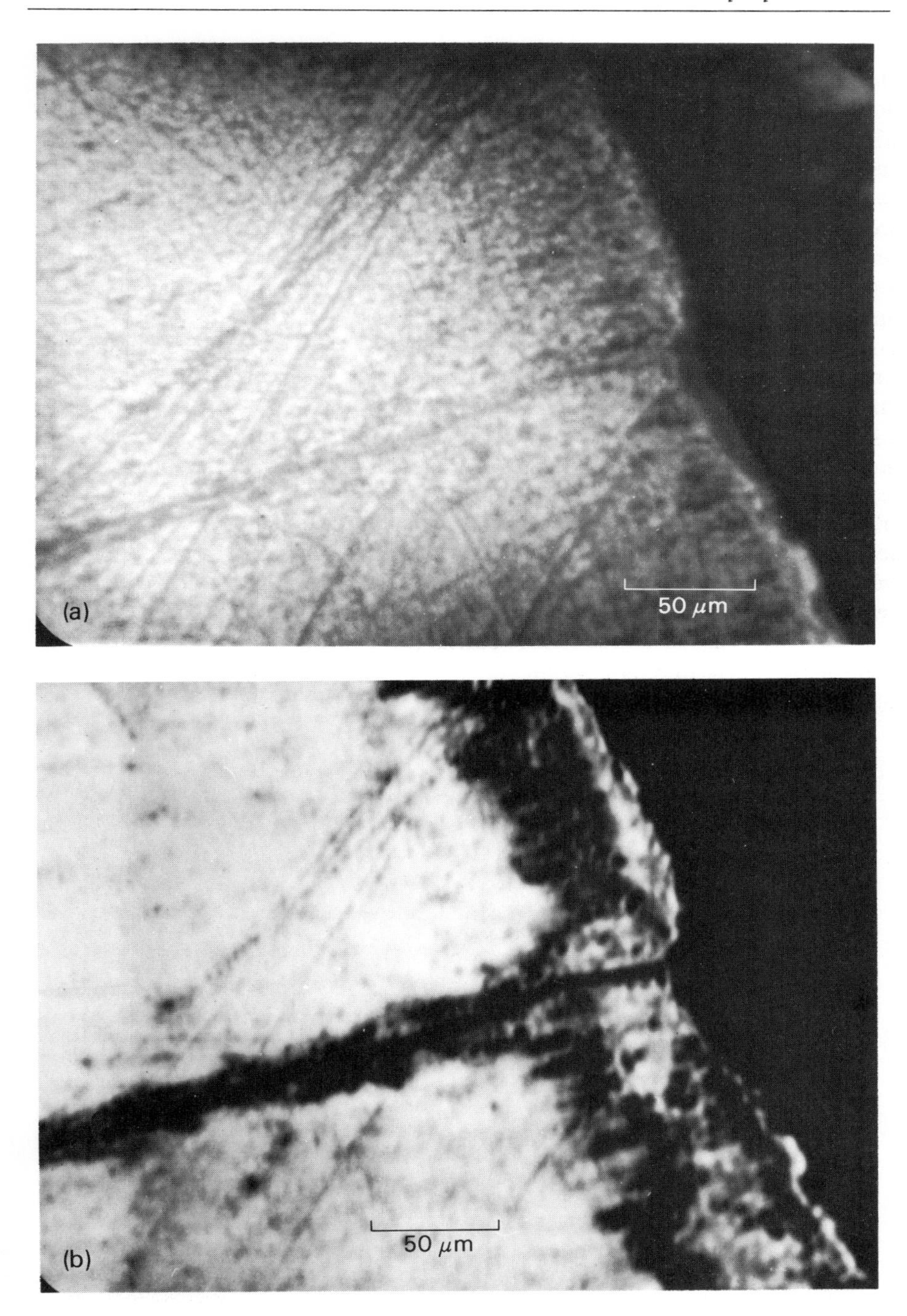

Fig. 4.6. Section of a human tooth: (a) optical; (b) acoustic (0.73 GHz).

acid environment in the mouth, the enamel is demineralized, and calcium and phosphorus are dissolved out of the apatite. Normally remineralization occurs later in the presence of saliva, but if the rate of remineralization is slower than the rate of demineralization, a white spot lesion of demineralized enamel can form. This appears white to the naked eye because more multiple scattering occurs in the demineralized region than in the relatively transparent enamel. In the light microscope (Fig. 4.6(a)), even the white spot lesion is too transparent to give much contrast, and little detail can be seen (although good contrast can be obtained in a polarizing light microscope in transmission through thin sections that have been imbibed in a suitable liquid). The acoustic microscope gives better contrast, indicating that the demineralized region has different elastic properties from healthy enamel; this correlates with measurements of other mechanical properties made using microhardness indentation. The lesion does not extend right to the surface; there is a thin layer of healthy enamel there. Demineralization is also evident along the crack in the enamel. It is important in preparing this kind of specimen to use a rigid metal lap to prevent differential polishing, otherwise the contrast might arise from variations in surface heights, which would be due to other mechanical properties such as strength. A good way to check this, apart from interference light microscopy, is to take an acoustic image at positive z, i.e. with the specimen beyond focus, where $V(z)$ depends only weakly on the specimen properties, so that if it is truly flat less contrast should be found than at defocus towards the lens.

Anisotropy

Grain structure is often very clearly and easily seen in the s.a.m. An example in stainless steel is shown in Fig. 5.1. The specimen was polished to a standard metallographic finish, using 1 μm diamond paste for the final stage, but was not etched. Thus in the optical microscope the surface looked uniformly bright. The acoustic images show the details of the grain structure, including a number of twins. The contrast depends sensitively on the amount of defocus (Atalar, Jipson, Koch, and Quate 1979). At focus little contrast was visible, the best image (a) being obtained at 4 μm defocus towards the lens. Increasing the defocus to 10 μm (b) causes radical changes in the contrast, and in some cases even reversal.

The contrast obtained from grain structure may be analysed by means of the Fourier optics model presented in Chapter 3. The problem is essentially one of determining the reflectance function to be used in eqn (3.6). The origin of the contrast is not, as with two-phase materials, because of variations in elastic properties. The grains have identical elastic properties, but these are anisotropic so that the stiffness tensor changes its orientation according to the crystallographic orientation of each grain. A consequence of this anisotropy is that it is now necessary to

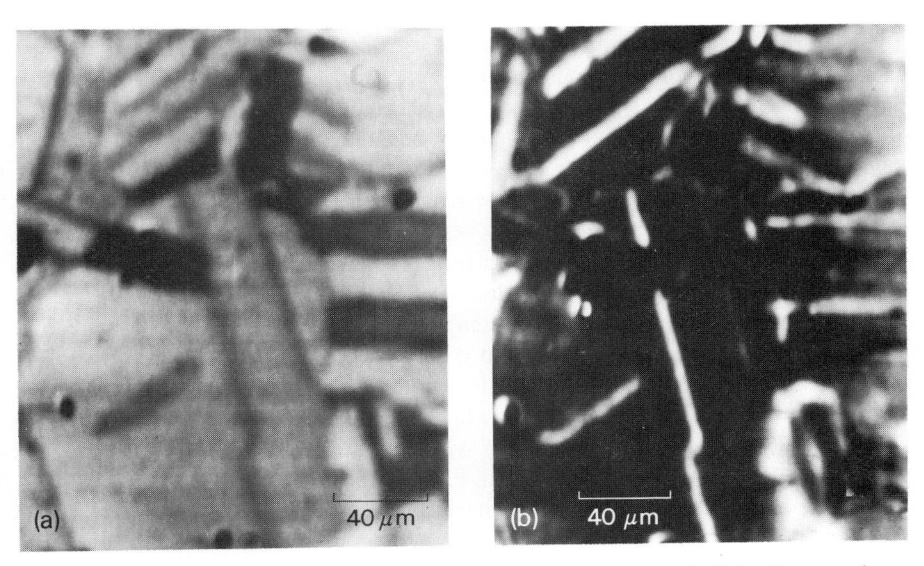

Fig. 5.1. Acoustic images of polished but unetched stainless steel (0.73 GHz): (a) $z \simeq -4$ μm; (b) $z \simeq -10$ μm.

specify not only the zenithal angle θ in the reflectance function, but also the azimuthal angle ϕ and the Miller indices $\{hkl\}$ of the surface, so that the reflectance function must be written $R_{\{hkl\}}\ (\theta, \phi)$.

For an isotropic material the reflectance function may be written analytically (eqns 3.7–10). For an anisotropic material this is not possible, and the following steps are necessary (Somekh, Briggs, and Ilett 1984). First the Christoffel equation is solved to compute the slowness surfaces. These are so called because they are surfaces in **k** space of the reciprocal velocity $\mathbf{k}/\omega$. There are three surfaces corresponding to the quasi-longitudinal wave and two quasi-shear polarizations. The slowness surfaces are rotated to correspond to a chosen crystallographic orientation, and Snell's law is solved for the liquid–solid interface by the requirement that the tangential component of **k** be continuous across the boundary. Some of the waves which satisfy this are discarded because they would give energy flow in the solid towards the liquid; they are identified by the orientation of the Poynting vector (the vector that describes power flow), which is normal to the slowness surface. Having selected three of the six solutions to Snell's law as physically allowable (though in some cases they may be imaginary), the acoustic Fresnel equations are finally solved by the requirement of continuity of traction at the interface (the tangential components being zero) and continuity of normal displacement. The reflectance function is then that part of the solution to the Fresnel equations that corresponds to the reflected wave in the liquid.

As an example of this, $R_{\{100\}}\ (\theta, \phi)$ for nickel is plotted in Fig. 5.2 for various values of ϕ. In many ways the first curve, along the [010] direction, $R_{\{100\}}\ (\theta, 0)$, is similar to a reflectance function for an isotropic material (c.f. Fig. 3.2). There are small variations in amplitude and phase at A, corresponding to the critical angle for propagation of longitudinal waves in the solid. Between A and B energy can propagate into the solid as a quasi-shear wave. At B the modulus of the reflectance function becomes unity, and beyond this angle no further energy can propagate into the solid. Rayleigh waves can be excited on the surface however, and this phenomenon leads to the phase change of 2π at C. It was this phase change which, when Fourier transformed, led to the oscillations with the periodicity of eqn (3.13). There is , however, a subtle difference between $R_{\{100\}}\ (\theta, 0)$ and an isotropic case. In the isotropic case, the slowness surfaces are spherical and so the critical angles occur when waves are being excited in the solid with angle of refraction of $\frac{1}{2}\pi$. The graphical solution of Snell's law for the anisotropic case is illustrated in Fig. 5.3. The angle to excite a shear wave parallel to the interface is θ_s, and this corresponds to point D in Fig. 5.2(a), where there is a scarcely visible fluctuation in $|R|$. But because of the shape of the slowness surface for quasi-shear waves with polarization in a vertical plane (SV), excitation can still occur up to the larger angle shown, and it is this angle which corresponds to B in Fig. 5.2(a).

Larger deviations from isotropic behaviour are observed as ϕ is increased. $R_{\{100\}}\ (\theta, 30°)$ is shown in Fig. 5.2(b). There are now two changes of 2π in the phase (note the change of scale for the phase of R). The phase change at $\theta \approx 34°$ corresponds to the excitation of a Rayleigh wave. The other phase change corresponds

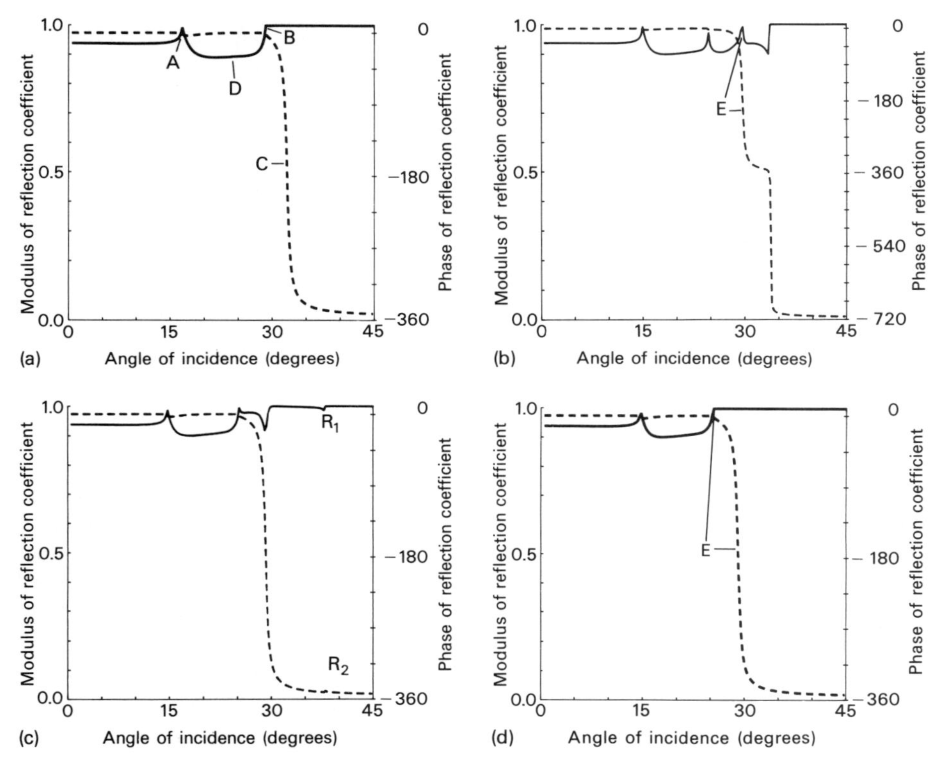

Fig. 5.2. The reflectance function of nickel for incidence on the {100} plane, for azimuthal angles (a) $\phi = 0$, (b) $\phi = 30°$ (the phase variation here is from $0°$ to $-720°$), (c) $\phi = 40°$, (d) $\phi = 45°$.

to the excitation of a pseudo-surface wave, a phenomenon which occurs only with anisotropic materials (Farnell 1970). This double phase change might be expected to have a marked effect on $V(z)$. At $\phi = 40°$ (Fig. 5.2(c)) the pseudo-surface wave (excited at $\theta \approx 29°$) dominates the reflectance function, so that the Rayleigh

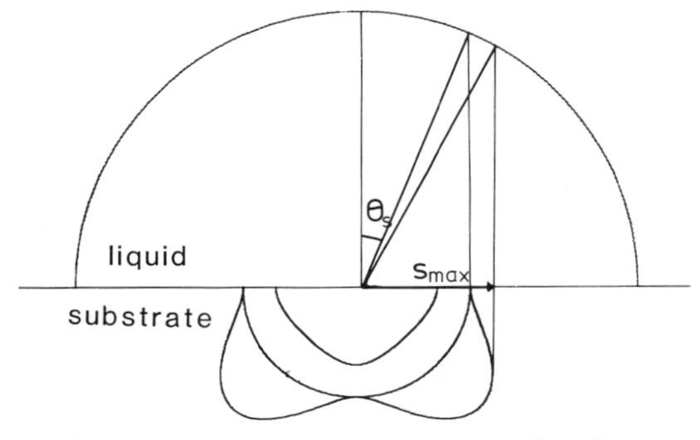

Fig. 5.3. Water–nickel slowness surfaces for incidence on the {100} plane; $\phi = 0°$. S_{max} denotes the largest tangential component of wavevector for excitation of quasi-shear waves.

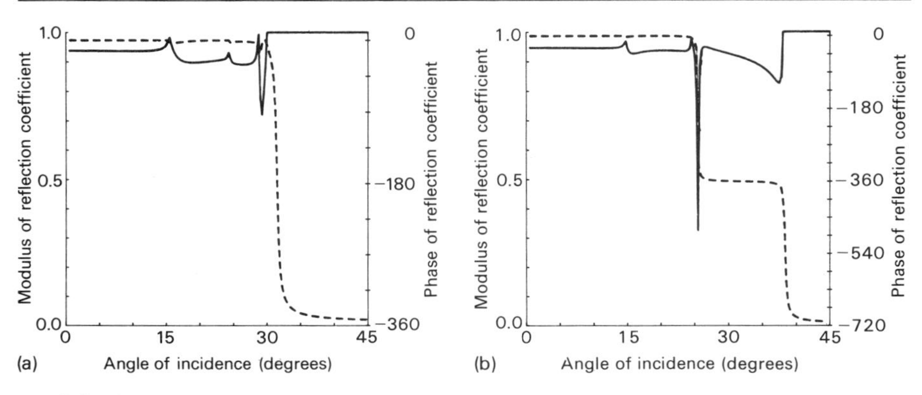

Fig. 5.4. Complex mean reflectance functions for the principal crystallographic planes of nickel: (a) $\{100\}$, (b) $\{111\}$.

wave (excited at $\theta \approx 38°$) introduces only small fluctuations (R_1 and R_2) in the reflectance function. Finally, for $\theta = 45°$ (Fig. 5.2.(d)), corresponding to incidence in the (011) plane, only one surface wave is apparent and the reflectance function again looks similar to that for an isotropic material.

In order to observe the $V(z)$ that would be produced by each of these reflectance functions a cylindrical lens may be used, thus making the situation essentially two-dimensional (Kushibiki, Okhubo, and Chubachi 1981). Oscillations in $V(z)$ are found which are in good agreement with eqn (3.13) when the appropriate angle for Rayleigh or pseudo-surface waves is used, and in particular $V(z)$ for a situation such as Fig. 5.2(b) shows two components of oscillation corresponding to each of the phase changes (Kushibiki, Horii, and Chubachi 1983). In order to compute $V(z)$ for a spherical lens, $R_{\{hkl\}}$ (θ, ϕ) must be integrated over all angles of ϕ to yield a complex mean reflectance function which can be used in eqn (3.6). Complex mean reflectance curves for the three principle crystallographic directions in nickel are shown in Fig. 5.4. Figure 5.4(a) is $R'_{\{100\}}$ (θ), and is the complex mean of functions of which examples were given in Fig. 5.2. It must be emphasized that no isotropic material could generate this function. This is even more apparent in Fig. 5.4(b), $R'_{\{111\}}$ (θ), in which features such as a dip in $|R|$ and a double phase change are manifested that are due to the shape of the slowness surfaces and the excitation of pseudo surface waves.

By using such complex mean reflection functions in equation 3.6, $V(z)$ may be calculated for various materials: some examples are presented in Fig. 5.5. Aluminium (a) is of low impedance and allows strong coupling to surface waves; there are therefore deep nulls (plotted here on a linear scale) in $V(z)$. However, aluminium has low crystal anisotropy so that differences in $V(z)$ for different orientations are small, at least as far as the second minimum. Copper (b) has a high anisotropy, but does not allow good coupling to surface waves, so that any differences between the orientations are small because the nulls themselves are small. Figure 5.5(a) and (b) thus corresponds to the experimentally observed facts that, for different reasons in

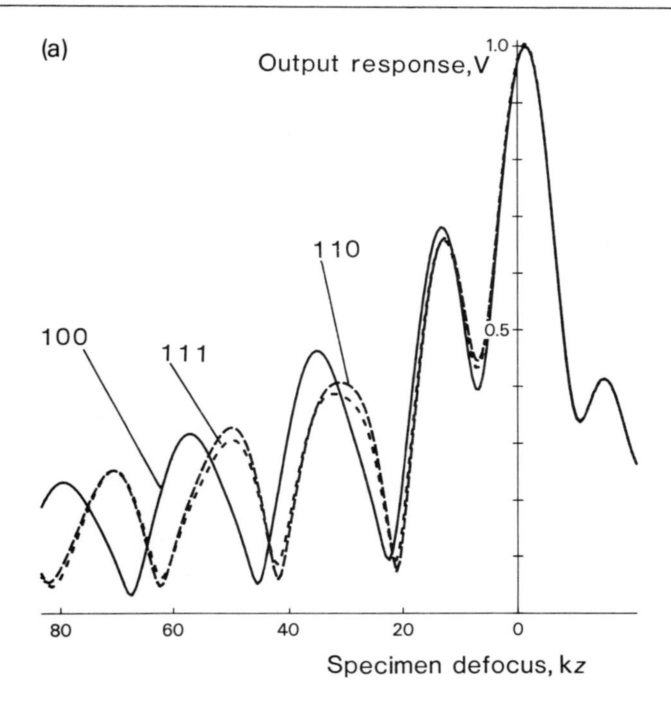
(a)
Output response, V
1.0
0.5
100
111
110
80
60
40
20
0
Specimen defocus, kz

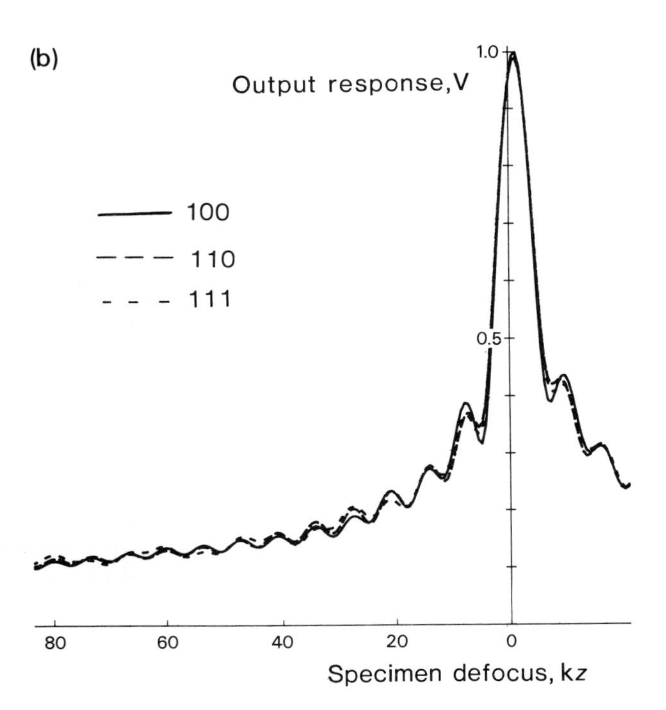
(b)
Output response, V
1.0
0.5
100
110
111
80
60
40
20
0
Specimen defocus, kz

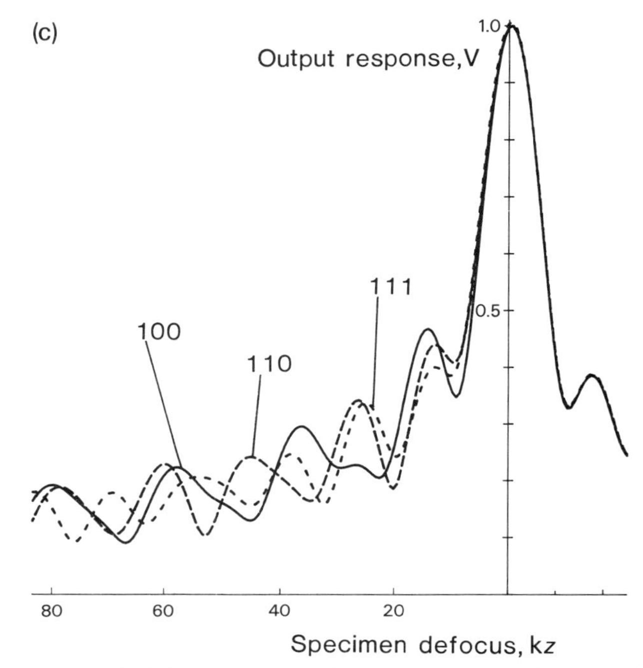

Fig. 5.5. (a) Calculated $V(z)$ for aluminium with a spherical lens of semi-angle $\alpha = 45°$; (b) copper, $\alpha = 60°$; (c) nickel, $\alpha = 45°$.

each case, it is not possible to obtain good images of grain structure in polycrystalline aluminium or copper using the acoustic microscope. So rather like Goldilocks, we find that aluminium is too isotropic, and copper is too slow, but nickel is just right!

Nickel, Fig. 5.5(c), is also expected to give little contrast at focus, or indeed when defocused by less than $kz \simeq -6$ (in water $k \simeq f/240\,\mathrm{s\,m^{-1}}$). But beyond this defocus, contrast begins to appear. Thus in a polycrystalline specimen of nickel at $kz = -9$, a grain of orientation $\{100\}$ appears relatively dark. From $kz = -14$ to -19 it will appear bright compared with other orientations; around $kz = -25$ it will be dark again, and so on. The micrographs of nickel in Fig. 5.6 indicate how such contrast reversals appear in practice, and help to show how practical images relate to $V(z)$. Figure 5.6(a) is a schematic sketch of the grains in the specimen, with three large grains labeled α, β, γ. A–B is a line along which line scans of intensity were measured that are presented below each micrograph (these are important because contrast and brightness were optimized electronically for each micrograph; the line scans enable absolute intensities to be observed and compared). Figure 5.6(b) was taken at focus: little contrast is observed. Figure 5.6(c) was at a defocus $z = -4\,\mu\mathrm{m}$; grain β is darker than α and γ. Figure 5.6(d) was at $z = -7\,\mu\mathrm{m}$, β is now brighter than α and γ; the contrast has reversed relative to $z = -4\,\mu\mathrm{m}$, and

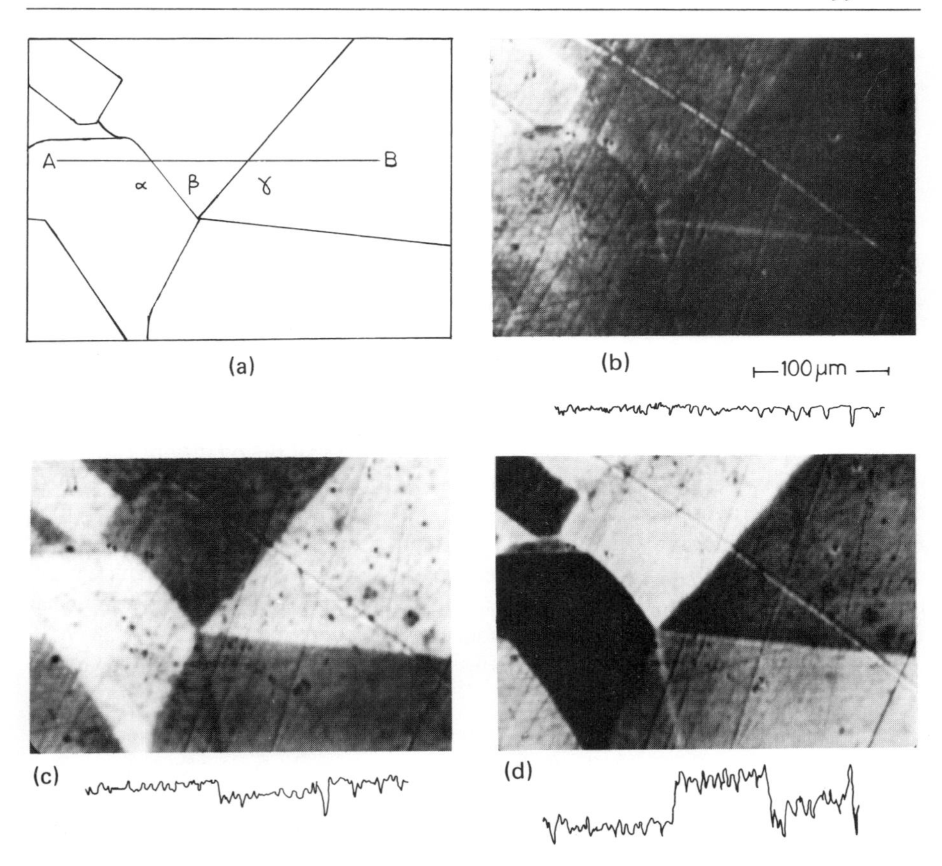

Fig. 5.6. Unetched polycrystalline nickel: (a) schematic diagram of grain structure; acoustic micrographs (0.73 GHz) and line traces along A–B at (b) $z = 0$, (c) $z = -4\ \mu m$, (d) $z = -7\ \mu m$.

the line scan indicates that the difference in intensities between grains is now considerably greater. Of course it is most unlikely that all of these grains had simple crystallographic orientations; nevertheless, it may help the reader to clarify the relationship between micrographs and $V(z)$ to follow these contrast reversals on Fig. 5.5(c) supposing that the grain orientations were, say, $\alpha\{110\}$, $\beta\{100\}$ and $\gamma\{111\}$.

6

Thin films and attenuation

When the reflection acoustic microscope was first demonstrated, an obvious application was to the inspection of integrated circuits. These are of enormous technological and commercial importance, and they come in a form ideally suited to examination in the s.a.m. They have surfaces that are already flat, so that little or no specimen preparation is required, and the films on the surface are sufficiently thin that they can be probed by the Rayleigh wave depth. An example of an acoustic image of an i.c. is shown in Fig. 6.1 (Miller 1983); the quality of the image is quite remarkable and compares favourably with anything that can be obtained optically. Acoustic micrographs of bipolar transistors on a silicon i.c. are shown in Fig. 6.2 (Hadimioglu and Quate 1983). The aluminium lines seen in Fig. 6.2(a) are about 3 μm wide. Another acoustic image of part of this area at higher magnification is shown in Fig. 6.2(b). These pictures were taken at 4.2 GHz, using non-linear properties of the water to give super-resolution by generating harmonics with half the fundamental wavelength at the focus of the acoustic beam. The resolution

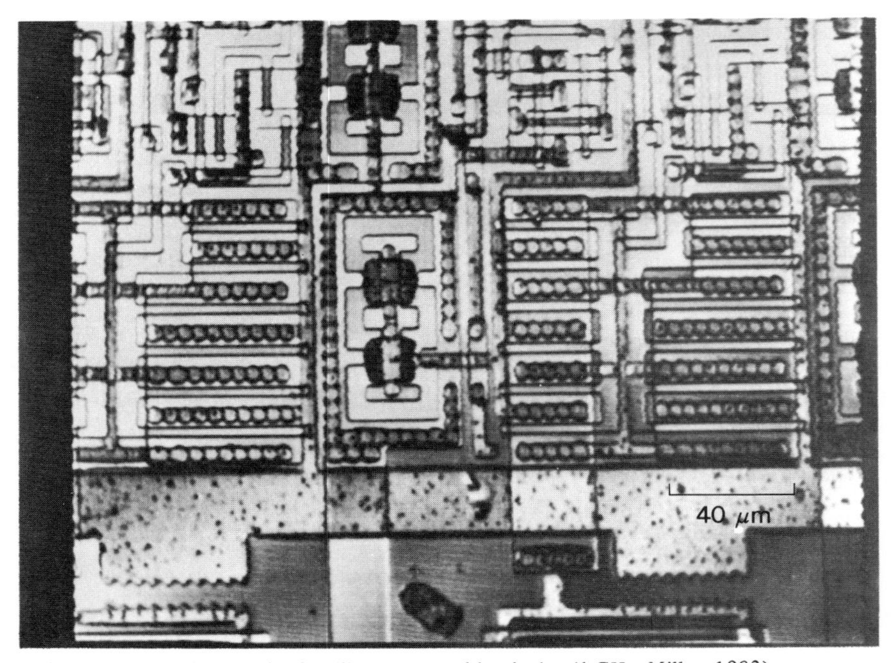

Fig. 6.1. Acoustic micrograph of a silicon-on-sapphire device (1 GHz; Miller, 1983).

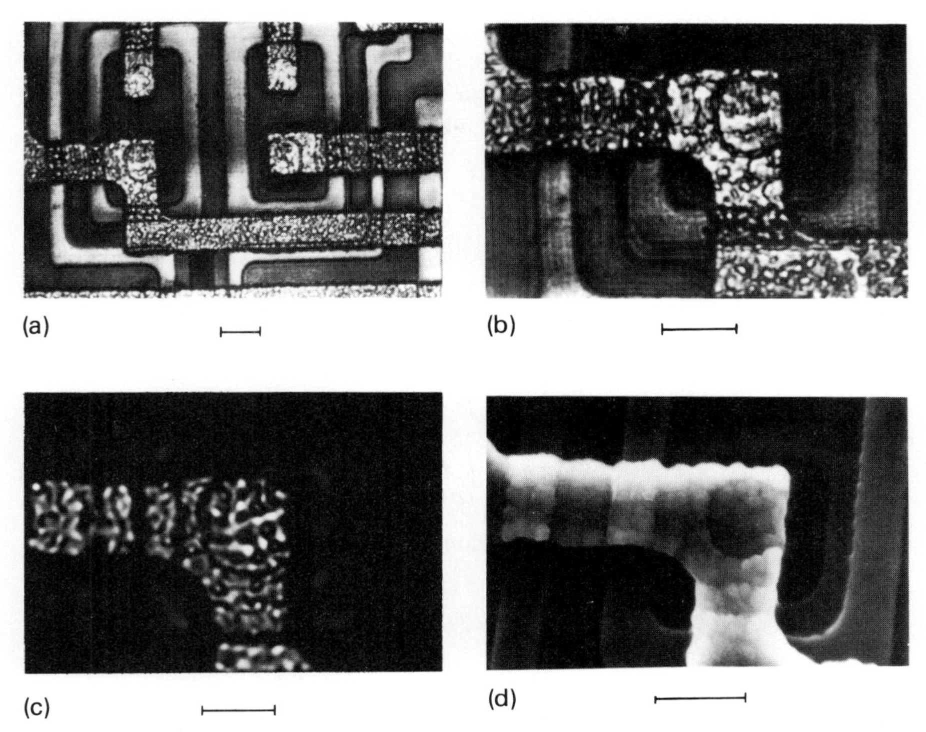

Fig. 6.2. A bipolar transistor on a silicon integrated circuit: (a) acoustic image (4.2 GHz); (b) higher magnification acoustic image of the base contact of a transistor; (c) optical and (d) s.e.m. micrographs of the same area as (b). Scale bars are 3 μm (Hadimioglu and Quate 1983).

achieved is $\leqslant 0.2\ \mu$m, and represents the state of the art for a water-coupled s.a.m. This resolution is at least as good as, and perhaps even better than, the resolution obtained in the optical micrograph in figure 6.2(c), which was taken with a 100 ×, 1.25 numerical aperture oil-immersion objective. The s.e.m. image in Fig. 6.2(d) offers higher resolution still, of course, and comparison between Figs. 6.2(b) and (d) indicates the extent to which the s.a.m. can image features that are more than simply topographical in nature.

In order to be able to interpret such images it is necessary to understand the effect of one or more surface layers on the reflectance function, and hence on $V(z)$. In Fig. 4.5 it was seen that a 0.1 μm layer of gold was transparent to the Rayleigh waves, in that the grain structure could be imaged through it, but it nevertheless affected the mean level of brightness over that region. The theory for calculating the reflectance function for a multilayered structure is given by Brekhovskikh (1980); for a full quantitative understanding of images of multilayered integrated circuits the method presented there must be used. It is not very difficult and lends itself nicely to implementation on a small computer. The case of a single layer that is thin compared with a Rayleigh wavelength may be approximately treated by considering the layer as a perturbation on the elastic properties of the substrate.

The fractional change in the Rayleigh wave velocity for an isotropic substrate with a thin isotropic layer of thickness h is (Auld 1973):

$$-\frac{\Delta V_R}{V_R} = \frac{V_R h}{4P_R}\left[\rho'|v_{Ry}|^2 + \left(\rho' - \frac{4\mu'}{V_R^2}\frac{\lambda'+\mu'}{\lambda'+2\mu'}\right)|v_{Rz}|^2\right]_{y=0} \tag{6.1}$$

where the primes denote properties of the overlay; $\lambda = c_{12}$ and $\mu = c_{44}$ are the Lamé constants and ρ is the density. The normalized particle velocity components in the substrate surface $y = 0$ with the wave propagating in the z direction are

$$\frac{(v_{Ry})_{y=0}}{P_R^{1/2}} = \left(\frac{f_y}{\rho V_s^2}\right)^{1/2}\omega^{1/2}$$

$$\frac{(v_{Rz})_{y=0}}{P_R^{1/2}} = \left(\frac{f_z}{\rho V_s^2}\right)^{1/2}\omega^{1/2} \tag{6.2}$$

where P_R = power flow per unit width along x, and

$$f_y = \left(\frac{V_s}{V_R}\right)\frac{4\eta^2(1-(V_R/V_s)^2)^{3/2}}{3\eta - 2\eta(V_R/V_s)^2 - 1}$$

$$f_z = \frac{1}{\eta}f_y$$

$$\eta^2 = \frac{1-(V_R/V_s)^2(V_s/V_1)^2}{1-(V_R/V_s)^2} \tag{6.3}$$

Once again, this formulation may be readily implemented on a computer. It does not give the perturbed reflectance function (for that the full treatment given in Brekhovskikh (1980) must be used), but it does enable the variation in the spacing of the minima in $V(z)$ to be related to the material properties and thickness of the film, by using the perturbed Rayleigh velocity. The Rayleigh velocity may be found from the periodicity in $V(z)$ from the expression

$$v_R = (1/\nu v_0 z - 1/(2\nu z)^2)^{-1/2} \tag{6.4}$$

where z is the spacing of the minima, ν is the frequency and v_0 is the velocity in the coupling fluid. $V(z)$ has been measured at 370 MHz for films on substrates (Weglein 1979*b*). For film thicknesses up to about 0.2 μm the spacing of the minima in $V(z)$ corresponds fairly well to the theory given above; for greater thicknesses it is necessary to use a more accurate model.

For examination of adhesion of films and coatings, it is often of more interest to obtain images than simply to measure $V(z)$. An example of this is given in Fig. 6.3 (Quate 1980). This shows optical and acoustic images of a chromium pattern deposited on a glass substrate. The optical image (a) simply shows the chromium and gives no information about adhesion. The acoustic images (b and c)

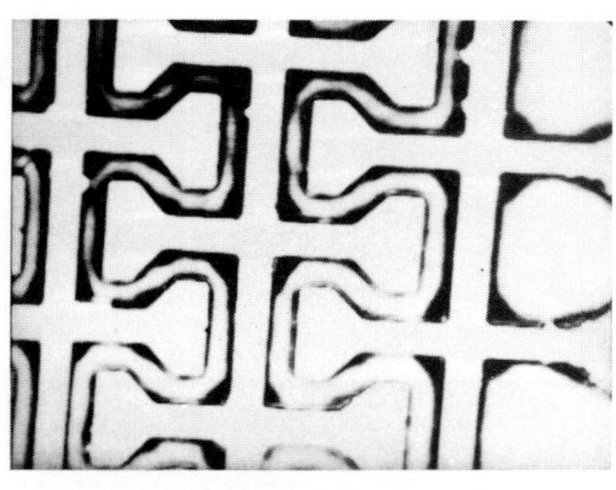

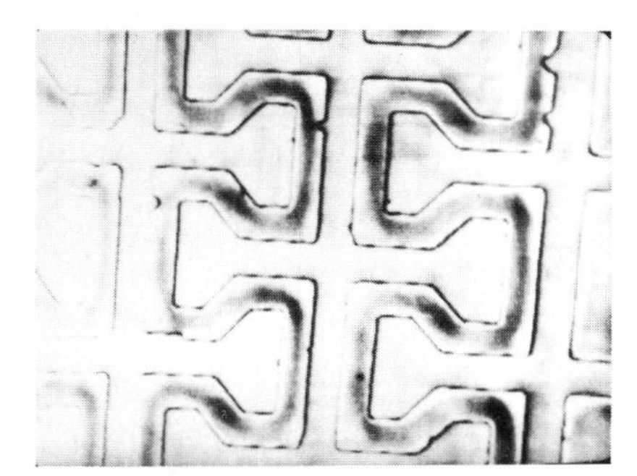

Fig. 6.3. 100 nm film of chrome on glass: (a) optical; acoustic (b) $z = -1$ μm, (c) $z = -0.5$ μm (1.5 GHz; Quate 1980).

show variations in the contrast of the chromium film, and subsequent destructive peal tests suggested that these correspond to variations in the quality of bonding to the substrate. Since the stresses associated with the acoustic wave are fairly small, the s.a.m. cannot reveal the strength of adhesion in a bonded region. However if a part of a film is actually detached from the substrate, then the reflectance function, and therefore $V(z)$, will be quite different from the case of the film on the substrate considered above. Therefore the variation of contrast with defocus will be quite different for bonded and detached regions of a film.

The layered structure of an integrated circuit device is more complicated, and therefore the acoustic images are even more interesting and potentially useful. Such images reveal the structure to the depth probed by the Rayleigh wave profile, and are sensitive to defects such as delamination within that depth. Figure 6.1 is an acoustic image of a silicon-on-sapphire finished device, and illustrates the wealth of detail potentially available from acoustic micrographs of integrated circuits (Miller

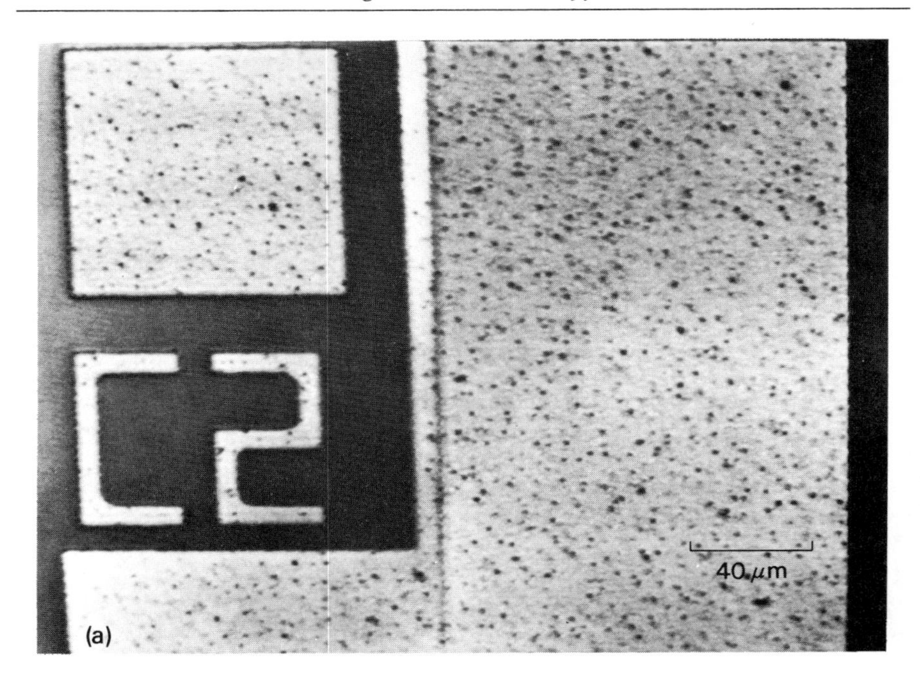

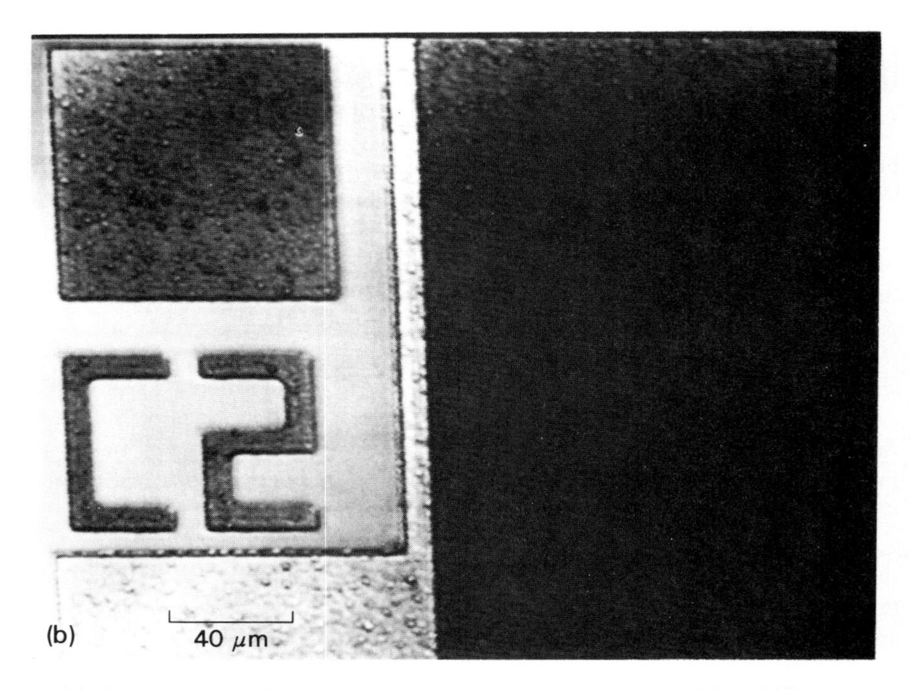

Fig. 6.4. Test capacitor on bulk silicon at two values of defocus (1 GHz; Miller 1983).

1983). The interpretation of such images, however, requires both considerable knowledge of the relevant contrast theory, and also as much *a priori* knowledge as as possible of the features and defects likely to be encountered. The cautionary notes that were sounded in regard to imaging theory generally apply especially here, with the added complication that an i.c. may have surface steps, so that in general different areas have different values of z. In a high resolution instrument, changing z must never be thought of as focusing below the surface (despite suggestions of this in the literature, e.g. Weglein 1983). Changing z has the effect of defocusing the reflection from the surface, which includes information from below the surface within the profile of a Rayleigh wave.

An example of the misinterpretation that can occur is given in Fig. 6.4 (Miller 1983). This shows a test capacitor on a silicon substrate. Figure 6.4(a) alone reveals the aluminium top surfaces, and suggests that they are fairly uniform. Figure 6.4(b), on the other hand, seems to show a great difference in the aluminium layer in the right-hand part of the picture. The difference in fact arises from a step due to a change in thickness of a buried vapox layer. Thus interpretation of either picture alone would be almost impossible, and in a case like this it is essential to study the behaviour of the contrast as a function of defocus. One way of presenting images at varying z is to use false colour, using a different colour for each value of z and then superimposing all the images (Hammer and Hollis 1982). While this may not often be practical it emphasises the need to interpret contrast in the light of its dependence on defocus.

This chapter began by considering the perturbation of the velocity of a Rayleigh wave by a layer on the surface. The Rayleigh wave can also be affected by attenuation in the material (Kushibiki, Ohkubo, and Chubachi 1982*b*). Such attenuation may occur either due to scattering, for example by grain boundaries, or by a loss mechanism, for example dislocation damping. The reflectance function can be directly calculated using the equations given in Chapter 3, and letting the longitudinal and shear velocities in the specimen surface have imaginary components corresponding to the attenuation per $\lambda/2\pi$. Theoretical curves of $R(\theta)$ for waves incident on aluminium (taken in this case as isotropic) are presented in Fig. 6.5 (Weaver, Briggs, and Somekh 1983). Figure 6.5(a) is for lossless aluminium, and shows the features that were described under isotropic contrast theory (cf. Fig. 3.2). Figure 6.5(b) is for an attenuation in both longitudinal and shear waves of 0.85 dB per wavelength. The phase change at the Rayleigh angle is almost identical to that in the lossless case, but the modulus shows a pronounced dip in precisely this region, so that the phase change will have a smaller influence on $V(z)$. Figure 6.5(c) is for a greater attenuation, of 4.5 dB per wavelength. The dip in $|R|$ is now less, though slightly broader, than in Fig. 6.5(b), but the phase change has reversed and is greatly reduced. Figure 6.5(d) is $V(z)$ for each of these three reflectance functions, computed for the same lens pupil function as Fig. 5.5(a). The effect of the increasing attenuation is first to reduce the average intensity, and then to reduce the amplitude of the oscillations in $V(z)$. The attenuation parameters for these figures were deliberately chosen to demonstrate the point nicely. Fortunately it

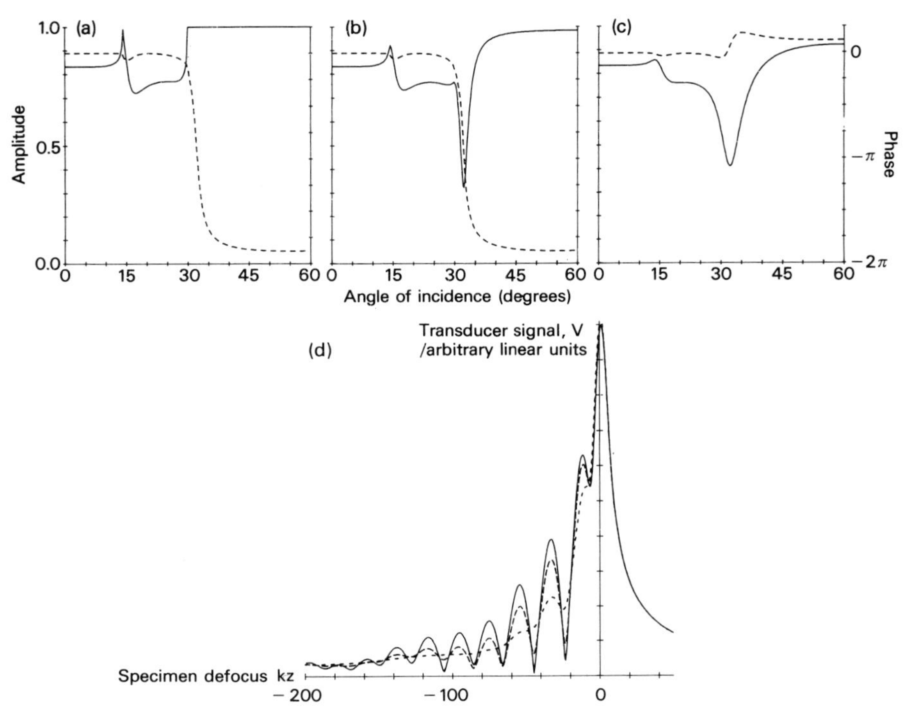

Fig. 6.5. The effect of attenuation on reflection from aluminium. (1) the amplitude and (2) the phase of the reflectance function for attenuations per wavelength of: (a) 0, (b) 0.85 dB, (c) 4.5 dB. (d) $V(z)$ for these reflectance functions [(a) ——, (b) – – –, (c) ---].

turns out that they can correspond well to those that are actually found in practice. Experiments with polycrystalline alloys using a s.a.m. at low frequency, so that the wavelength is longer than the grain size, show that the effect of grain size on $V(z)$ is quite noticeable (Yamanaka 1983). This means that in addition to the capability to image grains, described in the last chapter, it would be possible to work at a much lower resolution and image grain size itself, for example around the heat affected zone of a weld, the contrast arising now not from individual grains but from attenuation due to grain boundary scattering. Other experiments have been performed to measure $V(z)$ for a specimen of aluminium as it was deformed, this introduces dislocations and therefore causes attenuation by dislocation damping. Once again changes in $V(z)$ occur, so that it will be possible to image plastic deformation, for example at a stressed crack tip (Weaver *et al.* 1983). In both these cases non-destructive material characterization, measuring $V(z)$ but without imaging, would also be possible.

7

Cracks and discontinuities

Subsurface oblique cracks can generate interference fringes in acoustic microscopy. In Fig. 7.1(a) an acoustic image is shown of part of a quartz grain in granite. The fringes extend for a considerable distance to the left of the line where the crack breaks the surface, and their mean spacing is approximately 6.6 μm. Figure 7.1(b) is a dark field micrograph of the same area as Fig. 7.1(a). Since quartz is transparent the subsurface geometry of the cracks can be followed; the crack which is seen in the centre of Fig. 7.1(a) runs at an oblique angle to the surface of approximately 40°. Figure 7.1(b) also reveals that one of the cracks imaged in Fig. 7.1(a) is actually completely subsurface.

An explanation of the fringes observed in the acoustic images is illustrated in Fig. 7.2. It is assumed that two rays are important: ray A, which strikes the surface normally and is reflected back to the lens, and ray C, which is refracted at the specimen surface so as to strike the crack normally, and is also reflected back along its own path. As the lens is scanned parallel to the surface the path followed by ray A remains constant, but that of ray C varies (for scanning in the plane of the page). Since the transducer sums the amplitudes of the two contributions, periodic interference results, leading to the observed fringes. The periodicity is

$$\Delta x = \frac{\lambda}{2 \sin \theta_c}. \tag{7.1}$$

The exact velocity of sound in quartz depends on crystal orientation, a mean value is 5970 m s^{-1}. For $\theta_c = 40°$ this gives $\Delta x = 6.4\ \mu$m, which is consistent with the fringe spacing in Fig. 7.1(a).

A special case of the fringe pattern due to a crack in the specimen occurs when the crack runs normal to the surface, i.e. for $\theta = 90°$. Equation (7.1) then predicts a fringe periodicity of $\frac{1}{2}\lambda$. This proves to be the case, except that the wavelength is now not a bulk wavelength but the Rayleigh wavelength (Yamanaka and Enomoto 1982). An example is given in Fig. 7.3, which shows an acoustic image of cracks produced in a {111} surface of silicon by a microhardness indentor. The fringe patterns in the acoustic image have a spacing of 3.2 μm, giving a Rayleigh velocity of 4672 m s^{-1}, which is within the published range of 4540–4740 m s^{-1} (depending on orientation and direction of propagation; Slobodnik, Conway, and Delmonico (1973). Indeed, information about the crystallographic orientation of the specimen with respect to the indentor may be deduced from this measurement.) This mechanism also explains the fringes that are often seen in acoustic micrographs of semiconductor integrated circuits. These are present in Fig. 6.2(b); a rather more extreme case is presented in Fig. 7.4, where again the fringe spacing is $\frac{1}{2}\lambda_R$. These

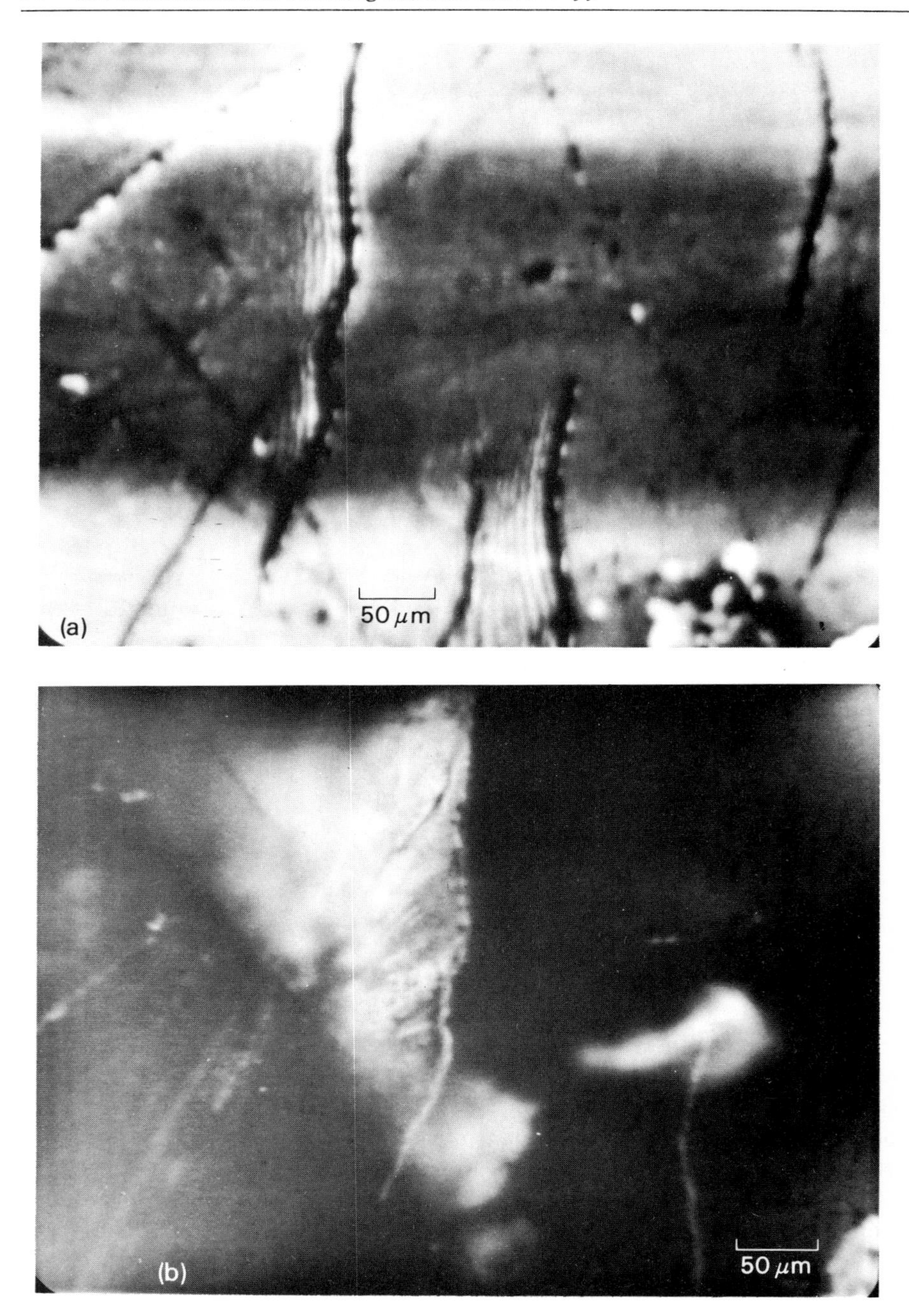

Fig. 7.1. Part of a grain of quartz in granite. (a) acoustic (0.73 GHz), (b) optical, dark field. The cracks that generate acoustic fringes run obliquely to the surface, and one of the cracks revealed in (a) does not reach the specimen surface.

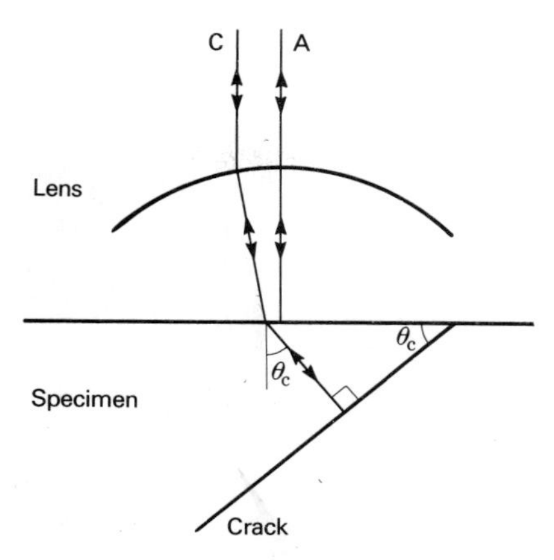

Fig. 7.2. Schematic diagram of the rays which are important in bulk wave interference fringes from angled cracks.

fringes can be eliminated by restricting the aperture of the lens so that it does not include the Rayleigh angle (Nikoonahad, Sivaprakasapillai, and Ash 1983), but of course this would be at the expense of much of the interesting contrast revealing features within a Rayleigh wavelength of the surface. These images emphasize the

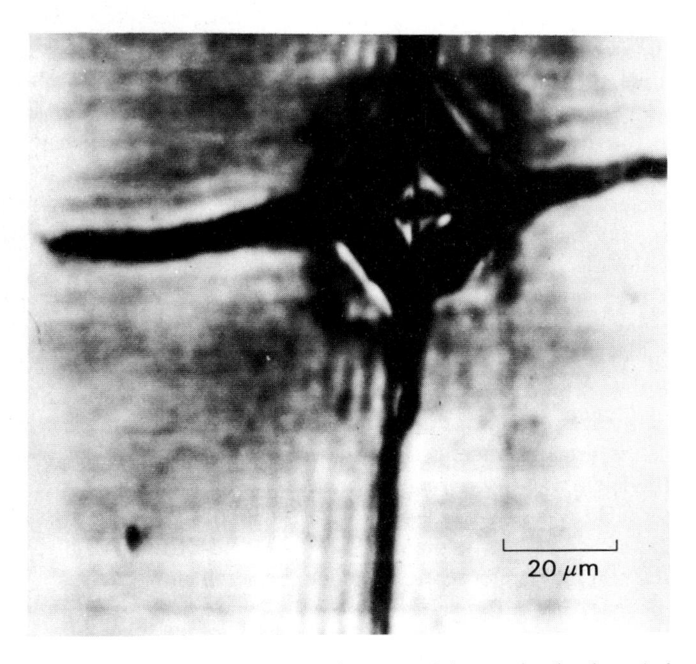

Fig. 7.3. Acoustic micrograph of a silicon surface containing a microhardness indentation. The fringe spacing is $\frac{1}{2}\lambda_R$ (0.73 GHz).

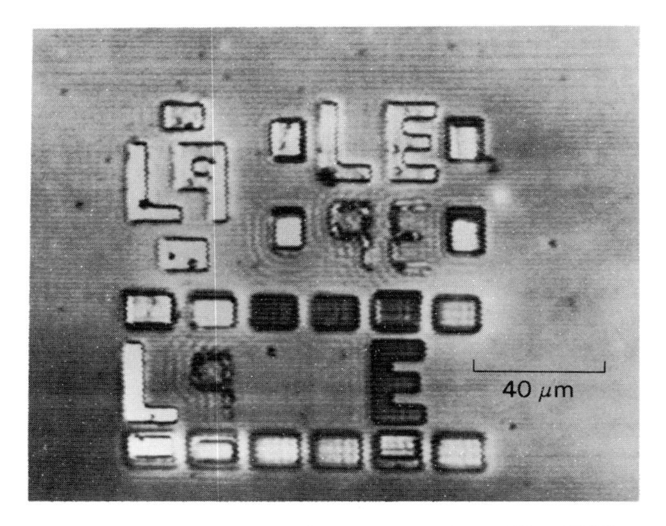

Fig. 7.4. Acoustic fringes observed on an SOS test insert (1 GHz; Miller, 1983).

importance of Rayleigh waves in the generation of the contrast observed in the acoustic microscope; indeed, although the $\frac{1}{2}\lambda_R$ spacing was presented as a special case of the interference it is actually the most common. Figure 7.1 is a further counter-example to the statement in Chapter 3 that in the high resolution s.a.m., contrast is not generally obtained from reflection by subsurface features. The reason here is that the subsurface echoes are not time-resolved; rather they interfere directly with the surface echo.

In many practical cases of surface cracks and other elastic discontinuities fringes are not seen, but nevertheless strong contrast may be found when the lens is directly over the discontinuity even though it may be much less than a wavelength wide (Ilett, Somekh, and Briggs 1984). This situation is illustrated in Fig. 7.5, in which the lens is over a crack. If the crack is thin, then waves incident close to the normal (beam A) will scarcely be affected by it, and will be reflected almost as if there were no crack. However, waves close to the Rayleigh angle (beam B) will excite Rayleigh waves, which may then strike the crack from the side. Such a wave will undergo strong reflection and scattering even when the crack is very thin.

When a Rayleigh wave is normally incident on a surface-breaking crack there is a reflection back along the surface of original propagation, transmission onto the second surface of the wedge or the crack surface, and diffraction into bulk waves. In the case of a crack of finite depth some energy is transmitted all round the surface to continue propagating in the original direction. In the presence of fluid there may be diffraction into the fluid and also transmission directly across the crack. It is assumed here that reflection along the original surface is the dominant effect (this would correspond to an infinitely deep crack filled with a vacuum). The reflection

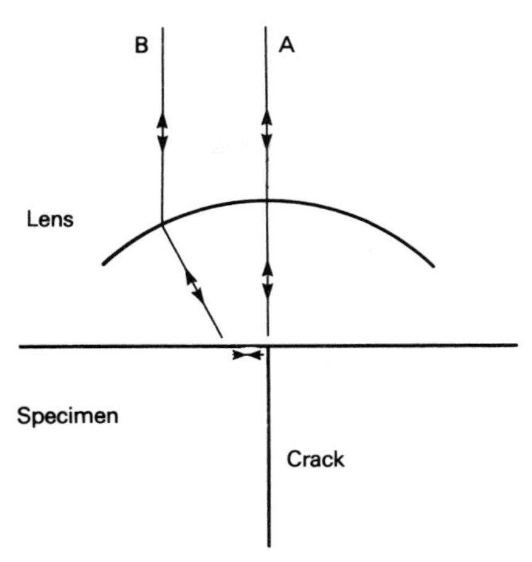

Fig. 7.5. The geometry when the lens is directly over a crack. Ray B is reflected by the crack and returns to the transducer along a path of the same length as it would have travelled in the absence of a crack.

involves a change of amplitude and phase which may be approximately described by a Rayleigh reflection coefficient (Achenbach, Gautesen, and Mendelsohn 1980):

$$R_R = 0.4e^{i0.6} \tag{7.2}$$

Since this result applies to Rayleigh waves incident normally on the crack, the lens in Fig. 7.5 is taken to be cylindrical, with the axis of the cylinder running parallel to the line of intersection of the crack with the surface. Because of the symmetry of the situation, rays which are reflected by the crack follow a path exactly equivalent to that which they would have followed had they been transmitted across the centre line in a defect-free specimen. Therefore the effect of the crack can be allowed for by multiplying that part of $R(\theta)$ which corresponds to Rayleigh wave phenomena by R_R (the effect at the transducer of re-radiation by Rayleigh waves before reflection by the crack is small and will be neglected).

Qualitatively, the result of this may be seen from the ray model: the effect of the crack on $V(z)$ will be to change the position and amplitude of the nulls, but not their periodicity. In order to make the approach quantitative it is necessary to isolate that component of the reflectance function which corresponds to excitation of Rayleigh waves. The reflection coefficient may be expressed as:

$$R(\theta) \equiv R_o(\theta) \left(\frac{k_x^2 - k_o^2}{k_x^2 - k_p^2}\right) \tag{7.3}$$

where k_x is the tangential component of the incident wave vector. k_o and k_p are determined by finding the imaginary incident wave vectors for which the numerator and denominator respectively of the analytical expression for $R(\theta)$ are zero. $R_o(\theta)$

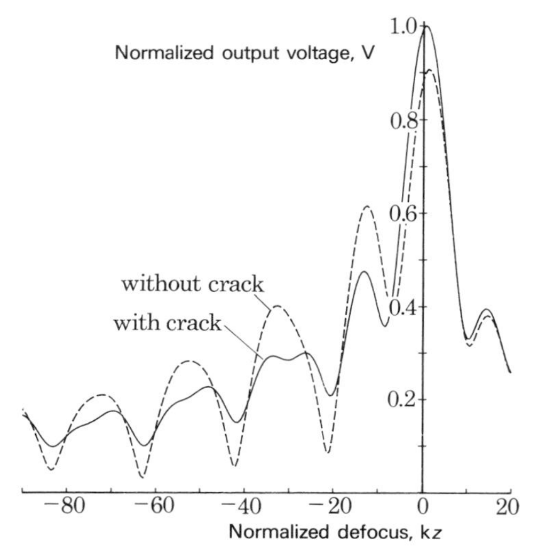

Fig. 7.6. $V(z)$ for the geometry of Fig. 7.7 generated using eqn (7.6): —— over a crack; – – – over a defect-free surface.

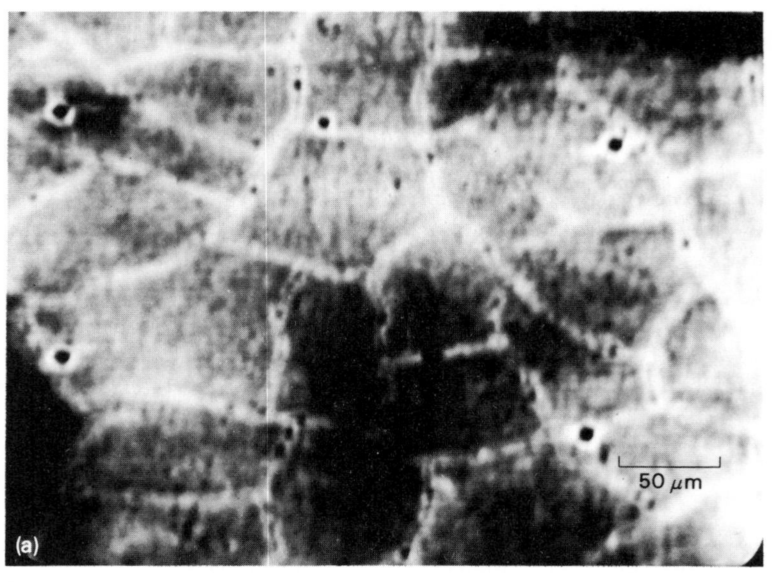

Fig. 7.7. The polished surface of a TiN coated hard metal cutting tool. (a) acoustic (0.73 GHz). The region around the top centre microhardness indent is shown at higher magnifications in (b) optical Nomaski and (c) s.e.m.

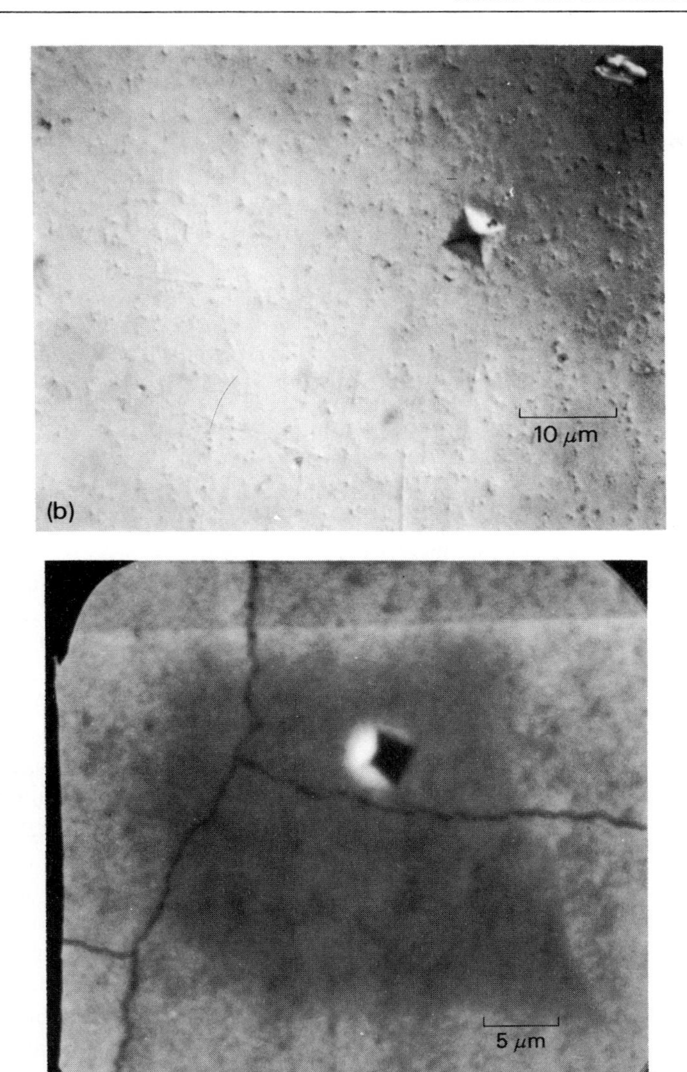

is defined by the identity and may be determined directly from it. The reflectance function may be separated into two terms:

$$R(\theta) \equiv R_o(\theta) + R_o(\theta)\left(\frac{k_p^2 - k_o^2}{k_x^2 - k_p^2}\right). \tag{7.4}$$

The second term on the right is associated with the singularity in the vicinity of $k_x = \pm k_P$, i.e. for waves incident around the Rayleigh angle. In the lossless case considered here $k_o = k_P^*$ (the asterisk denoting complex conjugate); the real part of k_P is approximately equal to the Rayleigh wave number for a free surface, and the imaginary part is due to re-radiation into the fluid. Thus the term describing phenomena associated with Rayleigh waves has been separated and for a cylindrical lens directly over a fine crack a local effective reflectance function may be written:

$$R_{\text{eff}}(\theta) = R_o(\theta)\left\{1 + R_R \cdot \frac{k_P^2 - k_o^2}{k_x^2 - k_P^2}\right\}. \qquad (7.5)$$

This expression may be inserted into the $V(z)$ formula for a cylindrical lens (Kushibiki, Ohkubo, and Chubachi 1982*a*) to give:

$$V_c(z) = \int_o^{\pi/2} P^*(\theta)\cdot R_{\text{eff}}(\theta)\cdot e^{-i2kz\cos\theta}\cdot\cos\theta\cdot d\theta \qquad (7.6)$$

This is plotted in Fig. 7.6 for a cylindrical lens of semiangle 45° (up to which $P(\theta)$ is taken as unity) over a crack in aluminium with water as the coupling fluid. For comparison $V(z)$ for a crack-free material is also plotted.

Images of a hard-metal cutting tool which has been coated with a film of TiN a few microns thick to improve wear resistance are presented in Fig. 7.7. The film was deposited by sputter coating at elevated temperature; subsequent differential contraction causes compressive thermal stress relief cracks to form in the coating. These cracks appear in the acoustic image (a) as bright lines. The width of the lines is about 8 μm and is determined primarily by the Rayleigh wavelength and the amount of defocus, and only weakly by the crack width. The acoustic contrast from a crack is distinct from that from a shallow surface scratch because Rayleigh waves can diffract around scratches with little perturbation. The optical image (b) was taken with Nomarski interference; the cracks are barely visible and cannot easily be distinguished from surface scratches. A scanning electron microscope image (c) confirms the pattern of cracks imaged acoustically; examination at still higher magnification revealed the cracks to be less than 200 nm wide at the surface (Ilett *et al.* 1984). Fatigue cracks are also revealed by this mechanism. Figure 7.8 shows optical (a) and acoustic (b) images of an Al–20 per cent Si plain bearing alloy that has failed in fatigue. The fatigue cracks are scarcely visible optically (though they could be made clearer by etching), but the acoustic image, in addition to revealing some of the microstructure of the specimen, reveals the cracks with considerably enhanced contrast.

Another kind of discontinuity which gives contrast is the boundary between two dissimilar materials. This is illustrated by the images of the glass fibre and epoxy composite material that were shown in Fig. 4.3. The fibres and the epoxy have different elastic properties and therefore give differing contrast, but the transition is not monotonic. Although the boundary must be very thin, it has a contrast of its own. In Fig. 4.3(b) it was bright; Fig. 4.3(c) was taken at increased defocus which caused contrast reversal, so that here the boundary appears dark. Finally, the point may be illustrated by the contrast from grain boundaries (Fig. 5.1). In the

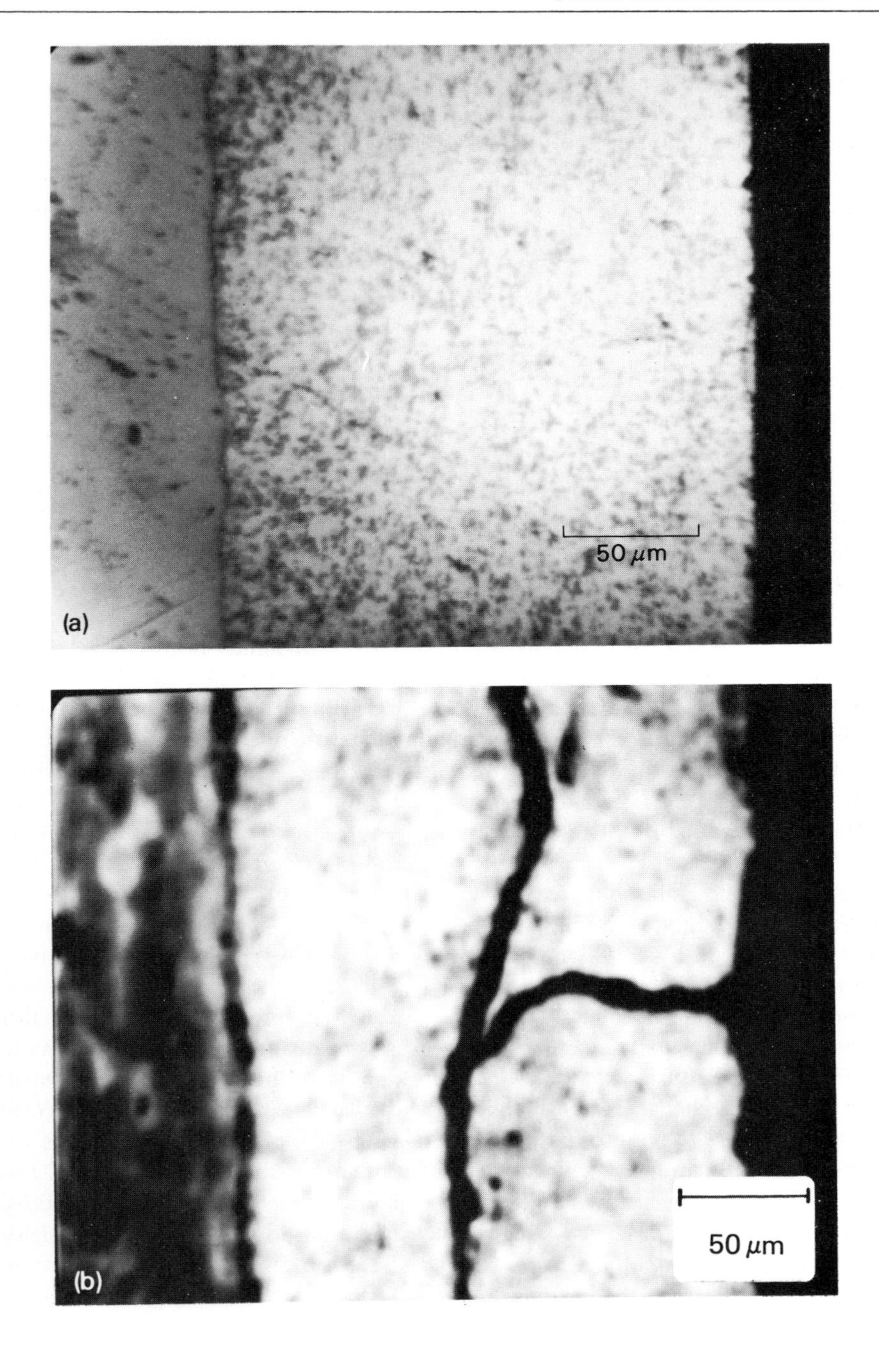

Fig. 7.8. An Al–20 per cent Si plain bearing alloy on a steel substrate (left). The optical (a) and acoustic (b; 0.73 GHz) micrographs are of the same area at the same magnification.

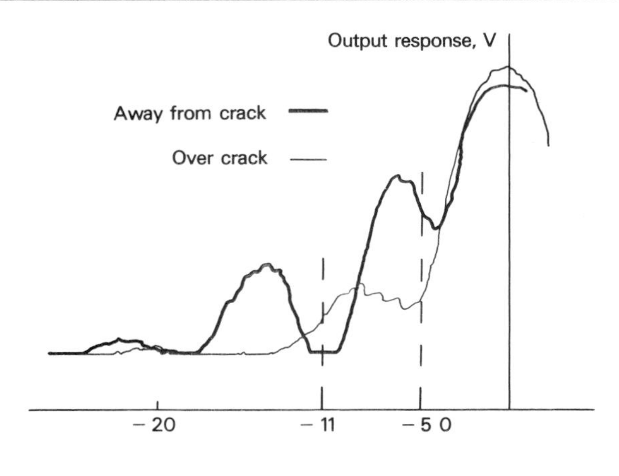

Fig. 7.9. Experimentally measured $V(z)$ for the Al Si bearing alloy. The indicated values of defocus correspond to the images of Fig. 7.10. (The origins of z for the two curves, and therefore their relative positions, are arbitrary).

case of the composite the boundary was between two dissimilar isotropic materials; here adjacent grains have identical anisotropic elastic properties but their elastic modulus tensors have different orientations. In addition to contrast between grain and grain there is contrast at the the grain boundaries themselves. Again, since these are only a few atoms thick this must be caused by the reflection of Rayleigh waves at the abrupt change in lattice orientation.

The distinctive nature of the contrast in the reflection scanning acoustic microscope is emphasized by the way it reverses with defocus, as illustrated by the appearance of the grain boundaries in Fig. 5.1. Reversal of contrast with defocus is a phenomenon exhibited by all the images presented in this chapter, and it may best be understood by consideration of $V(z)$. In Fig. 7.9 experimental $V(z)$ curves are shown for another region of the bearing alloy of Fig. 7.8. The lighter curve was measured with the lens directly over a crack, while the other curve is for the uniform surface far from the crack. Without pressing quantitative comparison with Fig. 7.6, the presence of the crack has the kind of influence that was suggested there, namely a change in the amplitude and phase of the oscillations in $V(z)$. Reversal of contrast corresponds to the changing over of the relative heights of the two curves. Figure 7.10(a, b, c) was taken of this region with a defocus of $z = 0, -5, -11\ \mu m$ respectively. In Fig. 7.10(a) poor contrast is seen, in Fig. 7.10(b) the crack appears dark against a light background, and in Fig. 7.10(c) it is bright against a dark background.

It thus appears that the enhanced contrast arises from the ability of an elastic surface to support Rayleigh waves. Because the acoustic microscope can couple into these, it is possible to image surface breaking cracks and other elastic discontinuities with waves which strike them from the side, and are therefore strongly affected

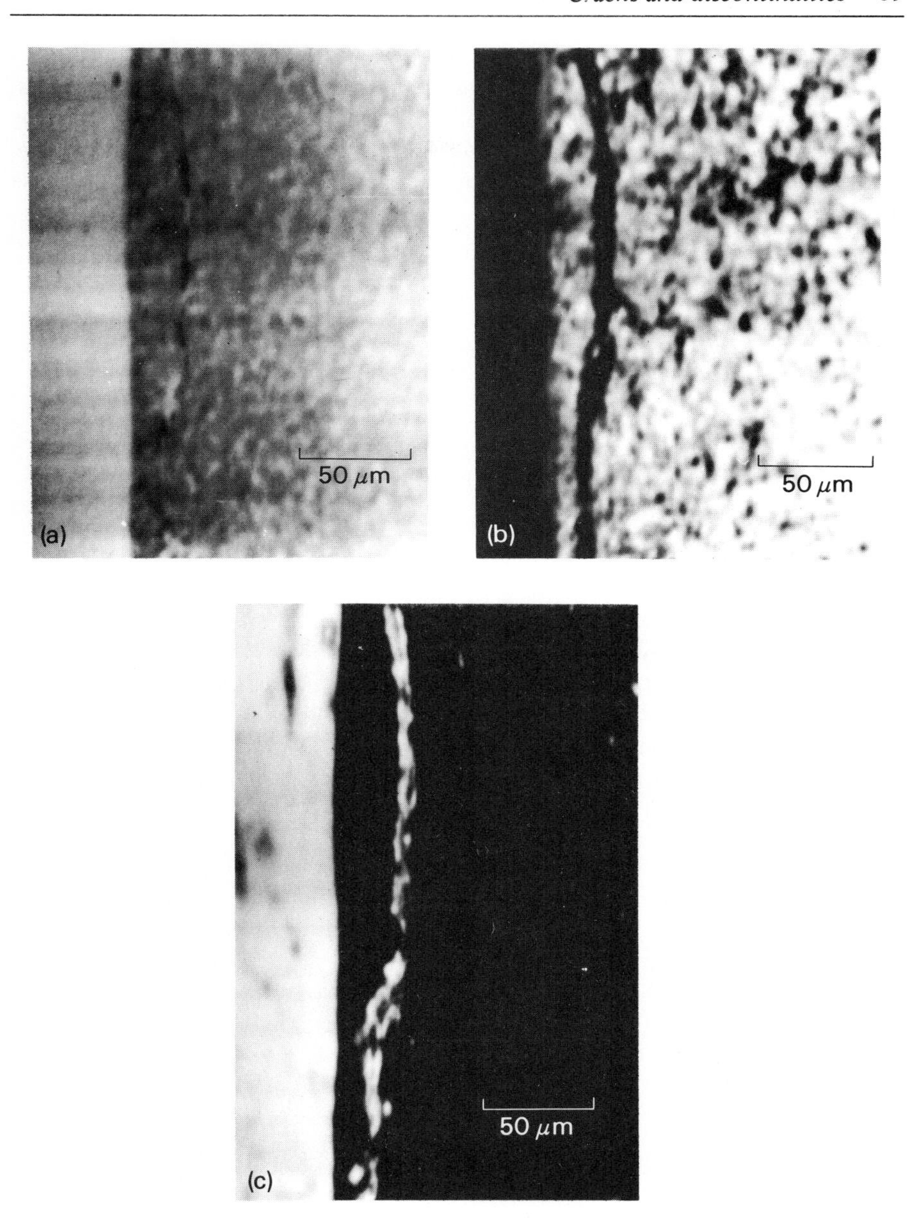

Fig. 7.10. Acoustic micrographs of another crack in the Al Si bearing alloy, at a defocus of (a) $z = 0$, (b) $z = -5\ \mu$m, (c) $z = -11\ \mu$m (0.73 GHz). Figure 7.9 was measured at two points on the line between the arrows, and the contrast between the crack and the alloy can be correlated with those curves.

even by features much less than a wavelength wide. By this mechanism, the reflection scanning acoustic microscope can give strong contrast from very fine defects. This contrast is available at much lower magnifications than would be necessary, for example, to detect the cracks using a scanning electron microscope, enabling large areas to be examined for such defects much more efficiently than would otherwise be possible.

8

Summary and conclusions

Rayleigh waves play a vital role in the contrast obtained in the scanning acoustic microscope; indeed for many purposes it is an accurate summary to say that the high resolution s.a.m. images the factors affecting the propagation of Rayleigh waves. The contrast theory may be thought of in two ways. In the diffraction model, it is never necessary to postulate the existence of leaky Rayleigh waves explicitly, because their behaviour is entirely analytic in the equations for the reflectance function that satisfy the boundary conditions for acoustic waves in a fluid incident on a solid surface. Such a reflectance function, when used in the $V(z)$ formalism, generates oscillations in $V(z)$ as found experimentally. But when one asks what the significance of these oscillations is, and what physical mechanism is responsible for them, the cause turns out to be the rapid change of phase of nearly 2π in the reflectance function at an angle of incidence slightly greater than the critical angle for shear wave excitation, and it is now well established that this in turn corresponds to the excitation of leaky Rayleigh waves in the surface of the solid (Bertoni and Tamir 1973). This justifies the use of a simpler and more intuitive model, in which it is explicitly assumed that the two rays of importance are the normally incident ray, and the ray that is incident at the Rayleigh angle, propagates in the specimen as a leaky Rayleigh wave, and then re-radiates a wave in the liquid symmetrical to the incident ray. This model predicts the same periodicity of oscillations in $V(z)$ as the Fourier model (and also accounts rather straightforwardly for the fact that these oscillations are found only for defocus towards the lens), and there is excellent agreement between the theoretical prediction, whichever way it is derived, and experiment.

In most high resolution s.a.m. images of the surface of flat specimens it is the Rayleigh wave propagation properties that are being imaged. The simplest cases are those in which each part of the surface can be regarded as an isotropic half-space. For typical solids the signal at focus is only weakly dependent on the elastic properties (specifically the acoustic impedance), and images are therefore taken at a small amount of defocus towards the lens. At this defocus the variation in $V(z)$ becomes rather sensitively dependent on the exact form of the reflectance function, in particular on the angle at which the 2π phase change occurs, and therefore contrast is obtained between regions of the specimen having different elastic properties. Because of the dependence of contrast on defocus, individual acoustic images can be ambiguous. It is not generally possible to interpret a single micrograph in terms of the elastic properties of the specimen; rather it is the way that the contrast varies with defocus that is important.

Few materials are truly isotropic on a microscopic scale. The reflectance function of an anisotropic surface cannot be written in closed form, but it can be found numerically. The most obvious effect of anisotropy is that the Rayleigh wave velocity now depends not only on the material, but also on the crystallographic orientation of the exposed surface, and this by itself would be enough to give contrast from grains in a polished but unetched surface. But there are two more effects. The first is that on a given surface the Rayleigh velocity also depends on its direction of propagation, so that different propagation directions tend to produce oscillations of different periodicity in $V(z)$. The second is that anisotropic surfaces can support pseudo-surface waves, so that for some directions there may actually be two changes of 2π in the phase of the reflectance function; this can lead to a second periodicity in the oscillations in $V(z)$, and in some circumstances this can beat with the oscillations due to Rayleigh waves. These effects can be nicely demonstrated by using a cylindrical lens that excites surface waves in one direction only. Usually, of course, imaging is performed with a spherical lens, and in that case these effects must be integrated around all angles. The resulting $V(z)$ is different from any that could be generated by an isotropic material, and because of the remaining dependence on the crystallographic orientation grain structure is often well revealed by imaging at a small amount of defocus.

Surface layers can affect the Rayleigh wave velocity, and can be considered as a perturbation on the elastic properties of the surface. For example, a layer of the same elastic moduli as the substrate but of higher density would slow the Rayleigh wave down, and to a first approximation the extent of the perturbation would be proportional to the thickness of the layer. After the effect of steps in surface height, this is almost certainly the dominant mechanism giving rise to contrast in acoustic micrographs of integrated circuits. Many of these have multilayers, and reflectance functions can be computed for these. Changing the defocus should be thought of in terms of $V(z)$ for such a reflectance function. From this, information can be deduced about features up to a wavelength below the surface, but defocusing must not be naively thought of as focusing below the surface. Similarly, contrast in images of debonded regions should be thought of as arising from changes in the velocity of waves in the surface. The propagation of Rayleigh waves can also be affected by factors that cause attenuation. These may be elastic scattering, such as that caused by microstructure on a scale smaller than the resolution of the lens, or by damping mechanisms such as dislocation damping. Since these reduce the contribution from Rayleigh waves at a given defocus, they affect the contrast, and it has been shown experimentally that both these kinds of attenuation can indeed produce noticeable changes in $V(z)$. Therefore it is possible to image in the s.a.m. effects that produce contrast by these mechanisms.

The reflectance function is defined only for a surface that is uniform, and therefore has translational symmetry. But some of the most interesting images on the s.a.m. come from discontinuities in surfaces, such as surface-breaking cracks, or boundaries between dissimilar materials or between grains. In this case a reflectance function cannot be written down, because scattering is present as well as specular reflection. The dominant role played by Rayleigh waves in all types of image considered so far gives the clue that is necessary to understand the enhanced contrast from surface discontinuities. If Rayleigh waves are excited in the surface, then instead of striking the discontinuity edge-on as waves in the fluid must do, and therefore being only very weakly scattered, they can strike it broadside, and therefore be strongly reflected. In the case of grain boundaries or boundaries between dissimilar materials, such reflection is due to impedance mismatch. In the case of a fluid-filled crack, since a Rayleigh wave is predominantly shear in character and fluids cannot support shear waves, very little energy is transmitted across the crack. By this mechanism surface cracks that cannot be seen with any other kind of microscope at the same magnification can be revealed with strong contrast in the acoustic microscope.

Many exciting developments are taking place in the techniques of acoustic microscopy. At least three areas can be identified. First, with every year that has passed since the invention of the scanning acoustic microscope, the available resolution has been improved. The most spectacular results are being achieved in helium below 200 mK at frequencies of up to 8 GHz, at which a resolution better than 25 nm is possible. Even in water, frequencies above 4 GHz have been successfully used, and by exploiting non-linear effects this has yielded a resolution better than that of an optical microscope. Second, there is tremendous potential for quantitative measurement using the acoustic microscope. By simple measurement of the periodicity in $V(z)$ the Rayleigh velocity can be deduced, and this has been achieved with accuracies of 0.2 per cent. By separating the pulses corresponding to normal and Rayleigh reflection and then employing a sensitive differential phase measurement system this accuracy can be enormously improved upon, and a potential resolution of one part in 10^5 may eventually be approached. Inversion techniques are being developed to obtain the reflectance function of the specimen, and if characterization of the lens and transducer can be successfully achieved then these will yield a great deal of information about its elastic properties on a microscopic scale, so that the microscope could be used as a quantitative elastic microprobe. Third, by using coupling fluids and specimens with reasonably compatible velocity and impedance, success is being obtained in imaging properties at some depth below the surface, thus exploiting the ability of ultrasonic waves to penetrate opaque materials.

No less significant than the developments that are taking place in the techniques of acoustic microscopy are those taking place in its application. There is a very important frequency range between what might be described as conventional ultrasonic non-destructive testing and the highest resolutions that are available, and Nikoonahad (1984) has identified subsurface imaging at 50–200 MHz as a leading application of imaging by the acoustic microscope. The higher frequencies are yielding important and exciting results in research in the field of materials science, and many more problems are waiting to be tackled. The scanning acoustic microscope is beginning to be applied to biological problems, and much important research can also be expected here. Several commercial scanning acoustic microscopes are now becoming available, and with the much wider access that this will bring to working instruments it will be possible for the use of the technique both for quality control and for research to become much more widespread. As this happens no doubt many new facets and abilities of the technique will emerge. It is hoped that this handbook will help that process.

References

Achenbach, J.D., Gautesen, A.K., and Mendelsohn, D. A. (1980). Ray analysis of surface–wave interaction with an edge crack. *IEEE Trans.* **SU-27**, 124–9.

Atalar, A. (1978). An angular spectrum approach to contrast in reflection acoustic microscopy. *J. appl. Phys.* **49**, 5130–9.

Atalar, A. (1979). A physical model for acoustic signatures. *J. appl. Phys.* **50**, 8237–9.

Atalar, A., Jipson, V., Koch, K, and Quate, C.F. (1979). Acoustic microscopy with microwave frequencies. *Ann Rev. Mater. Sci.* **9**, 255–81.

Attal, J. (1983). La microscopie acoustique. *Recherche* **14**, 664–7.

Auld, B.A. (1973). *Acoustic fields and waves in solids.* John Wiley and Sons, New York.

Bertoni, H.L. (1984). Ray optical evaluation of $V(z)$ in the reflection acoustic microscope *IEEE Trans.* **SU–31**, 105–16.

Bertoni, H.L. and Tamir, T. (1973). Unified theory of Rayleigh-angle phenomena for acoustic beams at liquid–solid interfaces. *Appl. Phys.* **2**, 157–72.

Brekhovskikh, L.M. (1980). *Waves in Layered media* (2nd edn). Academic Press, New York.

Briggs, G.A.D. (1984). Scanning electron acoustic microscopy and scanning acoustic microscopy: a favourable comparison. *Scanning electron Microsc.* **3**, 1041–52.

Derby, B., Briggs, G.A.D, and Wallach, E.R. (1983). Non-destructive testing and acoustic microscopy of diffusion bonds. *J. Mater. Sci.* **18**, 2345–53.

Farnell, G.W. (1970). Properties of elastic surface waves. In *Physical acoustics* (ed. W.P. Mason and R.N. Thurston) Vol. 6, pp. 109–66. Academic Press, New York.

Graff, K.F. (1981). A history of ultrasonics. In *Physcial acoustics* (ed. W.P. Mason and R.N. Thurston) Vol. 15, pp. 1–97. Academic Press, New York.

Hadimioglu, B. and Foster, J.S. (1984). Advances in superfluid helium acoustic microscopy. *J. Appl. Phys.* **56**, 1976–80.

Hadimioglu, B. and Quate, C.F. (1983). Water acoustic microscopy at suboptical wavelengths. *Appl. Phys. Lett.* **43**, 1006–7.

Hammer, R. and Hollis, R.L. (1982). Enhancing micrographs obtained with a scanning acoustic microscope using false-color encoding. *Appl. Phys. Lett.* **40**, 678–80.

Hildebrand, J.A., Liang, K., and Bennett, S.D. (1983). Fourier transform approach to materials characterization with the acoustic microscope. *J. Appl. Phys.* **54**, 7016–19.

Hildebrand, J.A. and Rugar, D. (1984). Measurement of cellular elastic properties by acoustic microscopy *J. Micosc.* **134**, 245–60.

Hollis, R.L., Hammer, R., and Al-Jaroudi, M.Y. (1984). Subsurface imaging of glass fibres in a polycarbonate composite by acoustic microscopy. *J. Mater. Sci.* **19**, 1897–903.

Ilett, C., Somekh, M.G., and Briggs, G.A.D. (1984). Acoustic microscopy of elastic discontinuities. *Proc. R. Soc. Lond.* **A393**, 171–83.

Jipson, V.B. (1979). Acoustic microscopy of interior planes. *Appl. Phys. Lett.* **35**, 385–87.

Kaye, G.W. and Laby, T.H. (1966). *Tables of physical and chemical constants* (13th edn). Longmans, London.

Kessler, L.W. and Yuhas, D.E. (1979). Acoustic microscopy – 1979. *Proc. IEEE* **67**, 526–36.

Kino, G.S. (1980). Fundamentals of scanning systems. In *Scanned image microscopy* (ed. E.A. Ash) pp. 1–21. Academic Press, London.

Kushibiki, J., Horii, K., and Chubachi, N. (1983). Velocity measurement of multiple leaky waves on germanium by line-focus-beam acoustic microscope using FFT. *Electron. Lett.* **19**, 404–5.

Kushibiki, J., Maehara, H., and Chubachi, N. (1981). Acoustic properties of evaporated chalcogenide glass films. *Electron. Lett.* **17**, 322–3.

Kushibiki, J., Ohkubo, A., and Chubachi, N. (1981). Linearly focused acoustic beams for acoustic microscopy. *Electron. Lett.* **17**, 520–2.

Kushibiki, J., Ohkubo, A., and Chubachi, N. (1982*a*). Theoretical analysis for $V(z)$ obtained by acoustic microscope with line-focus beam. *Electron. Lett.* **18**, 663–4.

Kushibiki, J., Ohkubo, A., and Chubachi, N. (1982*b*). Effect of leaky SAW parameters on $V(z)$ curves obtained by acoustic microscopy. *Electron. Lett.* **18**, 668–70.

Kushibiki, J., Sannomiya, T., and Chubachi, N. (1980). Performance of sputtered SiO_2 film as an acoustic antireflection coating at sapphire/water interface. *Electron. Lett.* **16**, 737–8.

Lemons, R.A. and Quate, C.F. (1974). Acoustic microscope – scanning version. *Appl. Phys. Lett.* **24**, 163–5.

Lemons, R.A. and Quate, C.F. (1979). Acoustic microscopy. In *Physical acoustics* (ed. W.P. Mason and R.N. Thurston) Vol. 14, pp. 1–92. Academic Press, London.

Miller, A.J. (1982). Aspects of SAM imaging of semiconductor devices. *Inst. Phys. Conf. Ser.* **67**, 393–8.

Nikoonahad, M. (1984). Recent advances in high resolution acoustic microscopy. *Contemp. Phys.* **25**, 129–58.

Nikoonahad, M., Sivaprakasapillai, P., and Ash, E.A. (1983). Rayleigh wave suppression in reflection acoustic microscopy. *Electron. Lett.* **19**, 906–8.

Nikoonahad, M., Yue, G.Q., and Ash, E.A. (1982). Subsurface broadband acoustic microscopy of solids using reduced aperture lenses. In *Review of progress in quantitative NDE* (ed. B.O. Thompson and D.E. Chimenti) Vol. 2B, pp. 1611–23. Plenum Press, New York.

Parmon, W. and Bertoni, H.L. (1979). Ray interpretation of the material signature in the acoustic microscope. *Electron. Lett.* **15**, 684–6.

Quate, C.F. (1980). Microwaves, acoustics and scanning microscopy. In *Scanned image microscopy* (ed. E.A. Ash) pp. 23–55. Academic Press, London.

Quate, C.F., Atalar, A., and Wickramasinghe, H.K. (1979). Acoustic microscopy with mechanical scanning – a review. *Proc. IEEE* **67**, 1092–114.

Sheppard, C.J.R. and Wilson, T. (1981). Effects of high angles of convergence on $V(z)$ in the scanning acoustic microscope. *Appl. Phys. Lett.* **38**, 858–9.

Slobodnik, A.J., Conway, E.D., and Delmonico, R.T. (1973). *Microwave acoustics handbook.* Volume 1A: *Surface wave velocities.* AFCRL-TR-73-0597, Air Force Cambridge Research Laboratories, Bedford, Mass.

Smith, I.R., Wickramasinghe, H.K., Farnell, G.W., and Jen, C.K. (1983). Confocal surface acoustic wave microscopy. *Appl. Phys. Lett.* **42**, 411–15.

Somekh, M.G., Briggs, G.A.D., and Ilett, C. (1984). The effect of anisotropy on contrast in the scanning acoustic microscope. *Phil. Mag.* **49**, 179–204.

Weaver, J.M.R., Briggs, G.A.D., and Somekh, M.G. (1983). Acoustic microscopy of ultrasonic attenuation. *J. Phys.* **12** (C9), 371–6.

Weglein, R.D. (1979*a*). A model for predicting acoustic material signatures. *Appl. Phys. Lett.* **34**, 179–81.

Weglein, R.D. (1979*b*). SAW dispersion and film-thickness measurement by acoustic microscopy. *Appl. Phys. Lett.* **35**, 215–17.

Weglein, R.D. (1983) . Integrated circuit inspection via acoustic microscopy. *IEEE Trans.* **SU-30**, 40–2.

Weglein, R.D. and Wilson, R.G. (1978). Characteristic material signatures by acoustic microscopy. *Electron. Lett.* **14**, 352–4.

Wickramasinghe, H.K. (1983). Scanning acoustic microscopy: a review. *J. Micros.* **129**, 63–73.

Yamanaka, K. (1983). Surface acoustic wave measurements using an impulsive converging beam. *J. appl. Phys.* **54**, 4323–9.

Yamanaka, K. and Enomoto, Y. (1982). Observation of surface cracks with scanning acoustic microscope. *J. appl. Phys.* **53**, 846–50.

Yin, Q.R., Ilett, C., and Briggs, G.A.D. (1982). Acoustic microscopy of ferroelectric ceramics. *J. Mater. Sci.* **17**, 2449–52.

Index